Telescopes in Space

Telescopes in Space

ZEDENEK KOPAL

Hart Publishing Company, Inc. • New York City

Telescopes in Space

Contents

Telescopes in Space

1 · The Story of Telescopes

Astronomy is one of the oldest sciences developed by the human mind; and a retrospective look at its age-long story reveals its course to be like a meandering river—long quiescent periods of slow gestation interrupted now and then by rapids when new vistas opened up suddenly to widen our horizons, usually thanks to new tools which the advance of technology placed in the astronomer's hands from time to time. One such landmark—which marks the beginning of our story—sprang out of a discovery which was to influence profoundly the whole subsequent development of astronomy and to enlarge vastly our horizons in space; namely, the invention of the telescope.

A story of the discovery of the telescope remains to some extent shrouded in a mystery which does not lack elements of the dramatic. It is, however, certain that towards the end of the first decade of the seventeenth century telescopes suddenly appeared in human hands in several places of western Europe—Middelburgh in Holland, Paris, Venice; but whether this appearance was due to independent discoveries or to rapidly spreading intelligence of the fact, is very difficult to fathom after a lapse of three and a half centuries.

This is all the more true as, from the very beginning, the telescope was not regarded primarily as an instrument of scientific research, but one which lent itself also to other eminently practical uses. For listen to what Galileo Galilei, one of the principals of this episode in the history of science, wrote to the Doge of Venice in August 1609 in commending to him this novel invention:

"The power of my cannocchiale to show distant objects as clearly as if they were near should give us an inestimable advantage in any military action on land or sea. At sea, we shall be able to spot the enemy warships and their flags two hours before they can see us; and when we have established the number and type of the enemy's craft, we shall be in a position to estimate his strength and decide whether to pursue and engage him in battle, or take flight. Similarly, on land it should be possible from elevated positions to observe the enemy camps and their fortification; and even in open field we should be able to see all his movements or preparations and follow them in detail."

And less than a year later, when Galileo was seeking to exchange the academic life of Padua for the Government service in Florence and was soliciting a position of the Principal Mathematician to the Grand Duke of Tuscany, this is what he wrote on May 7, 1610, to the Grand Duke's Prime Minister, Belisario Vinta:

"I have many and most admirable plans and devices; but they could only be put to work by princes because it is they who are able to carry on wars, build and defend fortresses, and for their regal sport make most splendid expenditures."

We do not perhaps need to continue the quotations to gather that Galileo Galilei was a shrewd man, wise in worldly affairs, who would be quite capable of drafting suitable research proposals to Defense Departments of any one of the Great Powers which displaced Venice and its contemporary rivals from the seven seas of this planet in our time. *Plus ca change, plus c'est la même chose.*

Was it the reported appearance of a telescope at the Rialto in the spring of 1609, or the news received by Galileo from Paris at about the same time, which drew his attention to the problem? Whatever happened, it is certain that soon thereafter he was able to construct superior tubes* of his own making; and when he turned them to the sky he made in a short time so many fundamental discoveries that modern observational astronomy was veritably born at Padua in the enchanted autumn and winter months of 1609–10 — just as modern theoretical astronomy was born at about the same time by the publication of Kepler's *Astronomia Nova* in Prague. It is these two milestones that mark the watershed between the ancient and modern astronomy; and it is only proper that our story should begin at that time.

Fuller details of these events have been told elsewhere; but we should like to add a few words concerning the optical qualities of these early telescopes. Students

---

* The name "telescope" for the new device does not seem to have been coined till later—probably during Galileo's second visit to Rome in 1611—and may be due to his friend Demisiani.

The telescopes of Galileo Galilei.

of the history of science will recall the amount of in-
credulity and opposition which greeted Galileo's early
telescopic discoveries. Was it due only to contemporary
inertia of thought or professional prejudice? Unqualified
answers in the affirmative often given by the historians
overlook, however, the fact that the telescopes in the
hands of Galileo's colleagues probably were—at least in
the earliest days of telescopic astronomy—very inferior
to those of his own. For, in spite of Galileo's assertion
that he invented the telescope "through deep study of
the theory of perspective", there is little doubt that his
process was essentially heuristic; and so were his methods
of producing the lenses. Needless to stress, there was by
1610 no established method for grinding and polishing
optical surfaces. Galileo appears to have succeeded in
this task better than his contemporaries, and scored his
observational triumphs as much by the skill of his hands
as by the fact that his mind was prepared to accept what
he saw.

Galileo is reported to have made many telescopes;
but knowing him as we do we shall probably not err in
surmising that he kept the best ones for his own use. In
a letter to Belisario Vinta of March 19, 1610, Galileo
reported that out of one hundred and more glasses which
he had ground "at great fatigue and expense" only ten
were ab. to show the newly discovered satellites of
Jupiter; and his best telescope, which Galileo called af-
fectionately the "Old Discoverer" (still preserved; see
page 16) magnified only thirty times. May these facts

not explain partly why his early telescopic discoveries
—mountains on the Moon, phases of Venus, or the
satellites of Jupiter which paved the way for acceptance
of the Copernican system—were confirmed by others so
slowly?

But apart from this, the great old man certainly deserves our full measure of admiration. It seems established that telescopes were in actual use in Holland and
France before Galileo ever saw one. They were, however,
apparently in the hands of diplomats and soldiers, who
had other uses for their spy-glasses than observing the
heavens. The Italian skies were not the only factor in
Galileo's favor!

The road of further development of astronomical
telescopes was rather circuitous and, in retrospect, re-
plete with misunderstandings. In the years following the
generation of Galileo, the principal obstacle to increased
telescopic power was seen in the chromatic aberration
of the objective, which increases with the curvature of its
surface. The easiest way to lessen it seemed to diminish
this curvature and increase the focal length. As a result,
the telescopes grew at first inordinately in length, to
usher the first geological age of optical Dinosaurs characterized by small heads on huge bodies. The apertures
of their objective seldom exceeded 6–8 inches; but their
focal ratios became extremely large, leading to focal
lengths in excess of those of most telescopes existing at
the present time. Thus the telescope with which Hevelius
at Danzig carried out most of his observations of the

Observing during the early years of the Paris Observatory.

lunar surface had a focal length of 158 ft. (see page 31).
Needless to say, telescopes of such great lengths could

The first reflector constructed by Isaac Newton. (Now in the possession of the Royal Society of London.)

be but crudely mounted. Astronomers of that time had mostly to dispense with any kind of a tubus (a series of diaphragms being a poor substitute for keeping away stray light), and the objective was often mounted at the end of a long pole, directed to different parts of the sky by means of ropes and pulleys (see page 19). Sometimes, in desperation, the astronomer dispensed with the mounting altogether, and fixed his objective to the roof of a building, waiting on the ground for a transit of his celestial object literally with an eyepiece in his hand (see page 25). No wonder that Hevelius (1611-87), under such circumstances, preferred the unaided eye for the measurement of stellar positions to the end of his life. Thus was the truly heroic age of observational astronomy —the age of Hevelius, Huyghens or the Cassinis—and their discoveries (rings and satellites of Saturn, motion and maps of the moon, etc.) are not seen in proper perspective until one considers the crude telescopic means at their disposal.

In the end, the long-necked telescopic Dinosauri of the second half of the seventeenth century vanished from the scene under their own weight as much as under the impact of new developments in astronomical optics which had taken place in the meantime; and one of them was a gradual introduction into practical use of the astronomical reflector. The idea of such an instrument was known already to Galileo Galilei (and described by him, through the mouth of Sagredo, in his *Dialogues on the Two Great World Systems*); but it was not translated

The 40 ft. telescope of William Herschel at Slough.

into practice until in 1671 by Isaac Newton. Newton's instrument—of a type still called after him, and in possession of the Royal Society of London today—(see page 20) was too small for active celestial exploration; its principal mirror was only 37 mm. in diameter and 16 cm. focal length and magnified thirty-eight times (a little more than Galileo's "Old Discoverer"). Even though modest in size, however, it translated for the first time into practice the germ of an idea conceived half a century before in Florence, which culminated in the 200-inch reflector of Palomar Mountain in our own time (see page 96); and may soon continue its career in space, to extend the domains of the observable Universe far beyond the limits imaginable to early pioneers of telescopic astronomy.

These developments were, however, slow to unroll; and many detours awaited on the way. Besides his positive contribution to astronomical optics in the form of the first reflecting telescope, Newton left our science also a negative legacy in the form of a mistaken assertion that it was impossible to achromatize a convex lens. The incorrectness of this assertion was, to be sure, proved in 1733 by Chester More Hall, and the first achromatic objective was actually produced by John Dollond around 1759. Such was, however, the weight of Newton's authority (as well as technical difficulties in producing achromatic objectives of larger size) throughout the eighteenth century that the pendulum continued to swing from dioptric to catoptric systems, and the stage was set for the first period of efflorescence of the astronomical reflector in the lifetime of William Herschel.

The story of this one-time musician and organist, who relatively late in life turned to astronomy to become one of the greatest observers of all times, has been told so often as to hardly call for a repetition in this place; but some remarks must be made concerning his achievements in astronomical optics. In the 1780s—at the time when Herschel turned from music to astronomy—there were as yet no professional telescope-makers to whom one could turn with an order for any but the smallest telescopes; and practicing astronomers were still very largely their own customers. Production of optical glass was still in its infancy; while the casting of metallic discs which could be polished into the shape of a mirror represented a much easier technical task—and one at which Herschel became singularly adept.

His *chef d'oeuvre* was the famous 20 ft. telescope of 18-inch free aperture and f/13 focal ratio (see page 26); and it was mainly with the aid of this excellent instrument that Herschel at last "coelorum perrupit claustra" and opened to his contemporaries, concerned still essentially with solar-system astronomy, the vistas of a much vaster world. His subsequent 40 ft. reflector (of 48-inch aperture), completed in 1789 (see page 22), was never a technical success; for a maintenance of the true optical form of its speculum mirror in all positions, or the mounting and control of so large an instrument confronted Herschel with problems which were insuperable

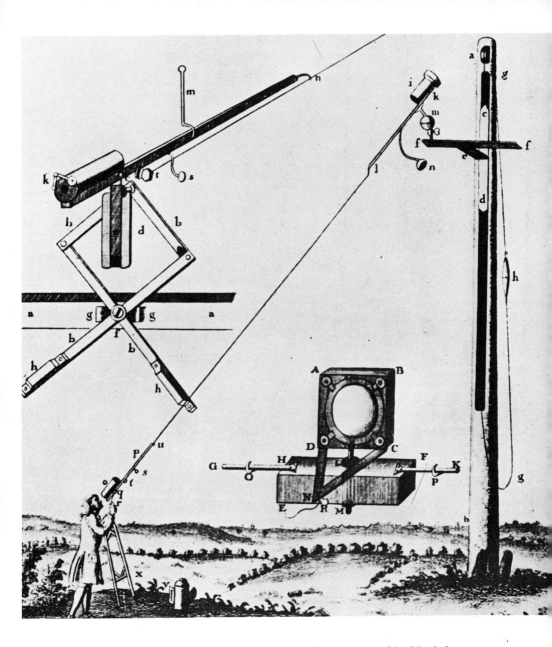

The aerial telescope of Christiaan Huyghens (second half of the seven-teenth century).

A drawing, made in the latter part of the eighteenth century, of the 20 foot telescope of 18-inch aperture with which William Herschel "pierced the barriers of the heavens."

in his time. Yet he was the first astronomer to use effectively the telescopic magnifying powers in excess of
1,000; and the optical quality of his mirrors was such
as to show stars "round as a button", to the astonishment
of the incredulous Henry Cavendish.

Herschel's instrumental triumphs between 1774 and
1823 inaugurated the first great era of astronomical reflectors which, however, also spent some of its momentum
with the decline of the ageing astronomer. It is true that
the achievements of his successors—such as his son Sir
John Herschel, W. Lassell or Lord Rosse—who eventually produced reflectors exceeding Herschel's in size
as well as performance, command our respect even today
(see pages 78-79). But, at the same time, new developments
in glass technology had opened up new possibilities for
the development of refracting telescopes. Thus Guinand
(1799) in Switzerland, Feil and others had at last
mastered the art of producing flint glass of the necessary
optical qualities; and these, in the hands of Fraunhofer
and his successors, had rapidly secured ascendancy to
the astronomical refractor, which was to last till the
end of the nineteenth century.

The Dorpat objective which founded Fraunhofer's
fame in 1824 was, to be sure, only 10 inches in diameter;
and still by 1865 the largest existing objectives (at Harvard and Pulkovo) possessed apertures no larger than
15 inches. Soon thereafter, however, refractors of ever-
increasing size began to come out from the hands of
Alvan Clark and his successors, culminating in the

40-inch Yerkes refractor in 1897 (see pages 81 & 165). Today, seventy years later, this refractor still remains the largest of its kind, as a witness of the fact that the evolution of its line more bogged down under its own weight—not that larger glass discs of requisite optical quality could not be produced (they actually were), but rather because increasing absorption of light in several inches of glass threatened to defeat their light-gathering power.

As a result, in the opening years of this century the pendulum of progress swung once more from refractors to reflectors: and the twilight of refracting telescopes went hand in hand with a new triumph of astronomical reflectors which continue to reign supreme in astronomy today, and whose fortunes—from the 60-inch reflector at Mt. Wilson in 1908 to the 200-inch telescope of Palomar Mountain dedicated to science in 1948 (see page 96)— owed much to the energy and vision of a single individual: George Ellery Hale. The relative ease with which optics for large reflectors could be produced, mounted and controlled has led, within our lifetime, to an unparalleled increase in celestial light-gathering power: for whereas by 1900 (i.e. at the end of the efflorescence of the refractors) the total light-gathering area of their objectives used for scientific work all over the world did not exceed 14,000 sq. inches, by 1960 this has increased to 247,000 sq. inches—an almost eighteenfold increase in sixty years—equivalent to a single 560-inch telescope. This increase in celestial light-gathering power is one of the principal reasons why we have been able to learn more astronomy from the sky in the past sixty years than all our ancestors found out in the preceding three centuries!

It is often said that progress in any branch of science or any other form of human endeavor, is largely controlled by the instrumental means which the current developments of technology have placed at the scientist's disposal; and the evolution of observational astronomy as sketched so far in this chapter bears out indeed such a contention to a remarkable extent. The discovery of the telescope and its impact on astronomy is very much a case in point; and so are the landmarks left by all the twists and turns of its evolution on the history of our subject. Thus the long line of astronomical refractors commenced with our ability to produce flint glass of the requisite properties and size; and the relatively slow progress of glass technology gave its first lease of life to astronomical reflectors towards the end of the eighteenth century. The reasons why this first efflorescence of the reflector proved to be relatively short-lived were again largely technological: namely, the inability to prevent rapid tarnishing of exposed optical surfaces (thus necessitating frequent polishing and refiguring of the mirrors), coupled with the inability to mount them in such a way that a relatively thin metallic mirror would not get disfigured by sagging under its own weight. The first of these drawbacks was, to be sure, lessened by the middle of the nineteenth century when Foucault introduced into astronomical practice glass mirrors with chemically silvered front surfaces—a practice which more recently was replaced by aluminization to give the reflecting coat a longer life-span; but the second had to await for more than a century following Herschel's time till the advances in glass technology have made it possible to produce large blanks of solid (or ribbed) glass or quartz, of sufficiently low thermal expansion, which are sufficiently free from internal stresses.

The reader may be interested to hear that, in recent years, the continuing advances in technology have made it possible to revive some of the earlier stages of the evolution of the telescope, and enable them to come into their own. The world's largest telescope of 236 inches of free aperture, now approaching completion in the U.S.S.R. (see page 208) will be of pyroceram coated with a special alloy to prevent tarnish, and supported by servomechanisms capable of a virtual elimination of flexures.

In the meantime, the techniques have also been perfected for grinding and polishing large optical surfaces to a high degree of precision (approximating the desired mathematical surface with less than one-tenth of a wavelength); while parallel advances in servo-control mechanisms have made it possible to automate telescope control well within the requirements of the observer. Does this mean that by the beginning of the last third of the twentieth century we have reached the stage at which the supporting branches of technology have over-

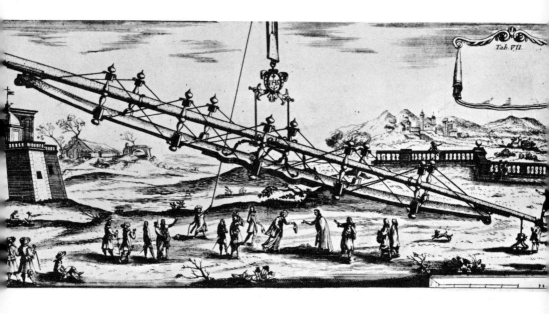

The long-focus telescope of Johannes Hevelius in Danzig.

taken the contemporary astronomical desiderata, and provided us with adequate technical means to undertake any astronomical work? The answer would indeed be almost in the affirmative—if it were not for one elusive but essential facet of the situation which bedevils the whole issue and which we have not mentioned so far in our narrative: namely, the atmosphere overhead and its effects on astronomical observations, which will be taken up in the next chapter.

2 · The Atmosphere Above Us

*If the Theory of making Telescopes could at length
be fully brought into Practice, yet there would be
certain Bounds beyond which Telescopes could not
perform. For the Air through which we look upon
the Stars is in a perpetual Tremor; as may be seen by
the tremulous Motion of Shadows cast from high
Towers, and by the twinkling of the fix'd Stars....*

*Long telescopes may cause Objects to appear
brighter and larger than short ones do, but these
cannot be so formed as to take away the confusion
of the Rays which arises from the Tremors of the
Atmosphere. The only Remedy is a most serene and
quiet Air, such as may perhaps be found on the tops
of the highest Mountains above the grosser Clouds.*

Isaac Newton, OPTICKS (4th ed., London 1730)
Book I, Part I, Propositio VIII.

Our description of astronomical telescopes as optical
systems imaging the scenery of the Universe around us,
as given in the preceding chapter, is still seriously amiss
in one respect: namely, it fails to take account of the fact
that observations from ground-based terrestrial observatories are being made at the bottom of an atmospheric
ocean. This ocean of air, consisting as it does of nitrogen
(78.08 per cent), oxygen (20.95 per cent), argon (0.93
per cent), carbon dioxide (0.03 per cent), variable
amounts of water vapor and other elements (neon,

33

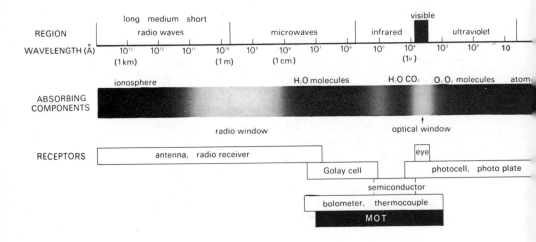

REGION

WAVELENGTH (Å)

long medium short
radio waves   microwaves   infrared   visible   ultraviolet

10^{13} (1 km) 10^{12} 10^{11} 10^{10} (1 m) 10^{9} 10^{8} (1 cm) 10^{7} 10^{6} 10^{5} 10^{4} (1μ) 10^{3} 10^{2} 10

ABSORBING COMPONENTS

ionosphere H_2O molecules H_2O CO_2 $O_2 O_2$ molecules   atom

radio window   optical window

RECEPTORS

antenna, radio receiver   eye

Golay cell   photocell, photo plate

semiconductor

bolometer, thermocouple

MOT

helium, krypton, xenon) or compounds (methane, nitric oxide, etc.) in amounts which the chemist would classify as impurities, extends to altitudes of several hundred miles, and exerts a pressure capable of balancing a column of mercury about 760 mm. in height. This pressure itself does not disturb us; for we all have become too accustomed to it to notice the burden; but the astronomer has special reasons to be irritated by it.

Before we explain them, let us preface this astronomical dislike of air with some qualifying remarks. That our planet borders on space through a gaseous fringe is no accident, but an inevitable consequence of the

LIGHT from celestial bodies passes through two discrete and widely separated "windows" in the otherwise opaque atmosphere. One of these windows is relatively narrow and exists in the visible region of the spectrum. Wavelengths from 2900A to 1u can pass through but all light below 2900A is absorbed by ozone molecules and light above 1u is absorbed by the combined action of water vapor and carbon dioxide. The second window extends in the radio region from wavelengths of 1 mm to several metres. Wavelengths greater than this are reflected by the ionosphere. Black rectangle at the borrom represents the range which manned orbiting telescope could encompass.

chemical composition of its mass, coupled with a gradual build-up of radiogenic heat in its interior. A degassing of this interior, by escape of thermally cracked volatile compounds, would be sufficient by itself to endow our planet with a gaseous envelope even if it initially had none; and the amount of gas which could be retained to form a permanent atmosphere depends, in turn, in a known way on the temperature and molecular weight of the constituent gases, as well as on the gravitational attraction of the planet's mass.

As a consequence of their large masses and low temperatures (due to their distance from the Sun), the

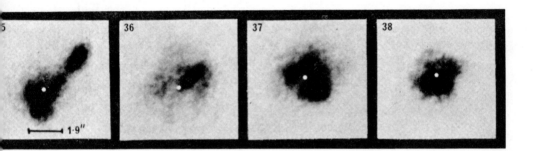

*Aldebaran: sequence of four images televised in rapid succession
to show effect of atmospheric disturbance.*

major planets of our system—Jupiter, Saturn, Uranus
and Neptune—were able to endow themselves with extensive atmospheres whose principal constituent is hydrogen, the lightest and cosmically most abundant of all
elements. The planets of the terrestrial size or mass—
Earth, Venus or Mars—possess atmospheres whose principal constituents are nitrogen and oxygen (forming
molecules substantially heavier than hydrogen). It is
only bodies of much smaller mass—such as the Moon

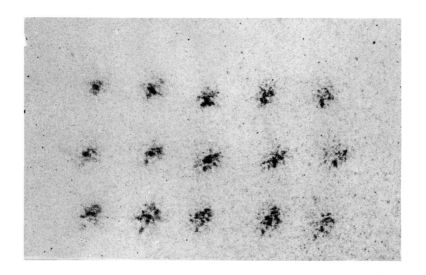

"Twinkle, twinkle little star" — in this case, Aldebaran — as photographed by J. Rosch with the Lallemand electronic camera at the Observatoire du Pic-du-Midi. Individual exposures of 2 milliseconds were taken at intervals of 25 milliseconds.

and other satellites of the solar system—that can afford to dispense with an atmospheric veil altogether; and these (for reasons to be explained later in this chapter) could be regarded as an astronomical approximation to a paradise. To be sure, astronomy—or, indeed, life— could not have developed on the surface of such a planet unprotected from the scorching radiation of the central star—at least not our kind of life (based on the biological carbon-nitrogen-oxygen cycle). It can, however, soon be exported there with adequate life-support; and, for an astronomer, the space-ships which are likely to take human beings to the Moon within our lifetime will offer a return journey to the paradise lost when our Earth began generating an atmosphere.

Why do the terrestrial astronomers feel so strong an urge to escape from its protective umbrella? The first and most obvious reason is the fact that its presence prevents astronomers from observing the stars in daytime (because of the scattering of sunlight on air molecules), and this deprives us of a half of the effective time when observations are possible from this globe (more than a half if twilight phenomena are considered). The scattering of sunlight on air molecules (depending as it does on wavelength) provides us in daytime with blue skies, too bright to see any but the brightest celestial objects through their veil; and limits the periods when the stars can be seen from the surface of the Earth to a few hours at a time. On the Moon, the surface of which is unprotected by any atmosphere to speak of, this would not be the case; the stars could be seen in the sky regardless of whether the Sun were above or below the horizon.

But, alas, skies are not always blue in daytime on Earth, nor are they clear at night as often as one would wish. Quite frequently (particularly in wet climates) all lights of the sky are blotted out by clouds of condensing water vapors; and their effects on astronomical observations are obvious beyond comment. In climates blessed with adequate moisture—such as that of Great Britain—more than 90 per cent of all nights are made unsuitable for astronomical observations through cloudiness (though more fortunate corners of the world are known where the opposite is the case).

Another annoyance to astronomical observation, for which the atmosphere is indirectly responsible, is caused by the light scattering on dust or other solid contaminants (produced by natural processes as well as human action), to which the air offers a convenient medium of support. Unlike gas molecules which scatter light selectively, the passage of light through dust whose grains are large in comparison with wavelength dims it non-selectively (i.e. reduces its intensity rather than altering its spectral distribution). Since, moreover, the ceiling of the dust layer hovers as a rule only a few miles above the surface of the Earth (as any air traveler can verify by direct observation during the ascent or descent of their aircraft), astronomers, following Newton's suggestion quoted at the head of this chapter, have grown accustomed to minimize its effects by taking their instruments to high mountains (see page 85)—a strategy which unfortunately, does not work quite so well with the clouds.

The atmospheric effects described so far could be classified as annoyances rather than as basic obstacles to astronomical work; and the main quality needed to defeat them would seem to be patience. Worse, however, is in store for us; for quite apart from the support

OVER

the air offers to solid or liquid suspensions, other atmospheric effects on astronomical observations are more insidious, and not necessarily limited to the troposphere.

In order to bring them in focus, let us recall that no telescopic system which astronomers use for their ends really commences with its objective or mirror used to collect light of any celestial source, but rather several miles above it—where the originally plane wave of light from space will commence to be distorted by a passage through turbulent atmospheric layers of unequally heated air of increasing density, and suffer anomalous refraction and diffraction from the turbulent elements long after it has undergone selective absorption very much higher up.

Let us explain the meaning of these terms in a few words. It may perhaps come as a surprise to some readers to learn that the atmosphere overhead—far from being transparent as it seems—is really opaque to most kinds of radiation (electromagnetic as well as corpuscular) incident upon it from space. The radiation which can penetrate to ground without hindrance is, in fact, limited to two discrete and well-separated domains of the spectrum (see page 34). The first is a relatively narrow "optical window" between wavelengths of approximately 2,900Å and 10,000Å—including the light visible to the human eye, which stretches from about 4,000Å to 7,000Å in wavelength. The fact that the "visible light" falls in the domain of atmospheric transparency is no accident; for the sensitivity of the human eye (or the

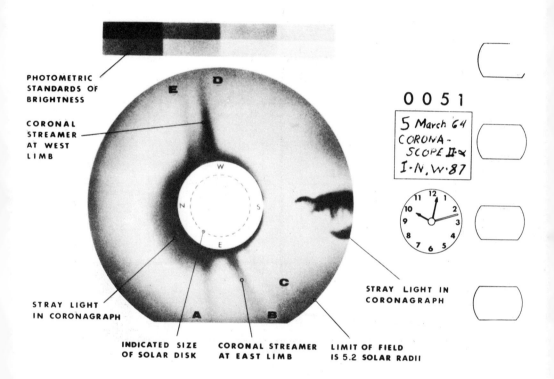

PHOTOMETRIC STANDARDS OF BRIGHTNESS

CORONAL STREAMER AT WEST LIMB

STRAY LIGHT IN CORONAGRAPH

INDICATED SIZE OF SOLAR DISK

CORONAL STREAMER AT EAST LIMB

LIMIT OF FIELD IS 5.2 SOLAR RADII

STRAY LIGHT IN CORONAGRAPH

0051

5 March '64
CORONA-
SCOPE II-α
I·N·W·87

A photograph of the solar corona taken with a coronagraph borne to an altitude of 80,000 ft. by means of a balloon (after Newkirk and Bohlin).

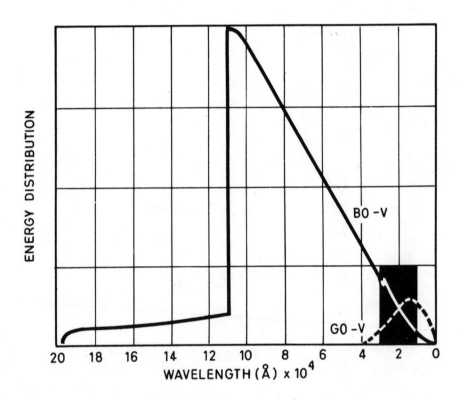

Graph showing distribution of energy on the spectrum of a star akin to that of the Sun, and of a younger star of early spectral type.

light receptors of other animals) have developed throughout the ages so as to take the best advantage of available
sunlight. The second "window" through which our atmosphere becomes transparent to give us another glimpse
of the Universe around us opens up in the radio-domain
of the electromagnetic spectrum—extending from wavelengths of approximately one millimeter up to several
meters.

What sets these limits on the atmospheric transparency
windows? The short-wave violet edge of the "optical
window" is imposed quite abruptly by the absorption of
the ozone (O_3) molecules, which in accordance with
the photochemistry of our atmosphere are formed in a
layer extending broadly from some 82,000 to 196,000
ft. above the surface, with the maximum concentration
fluctuating around 131,000 ft. Ozone constitutes the
principal source of absorption for ultraviolet light whose
wavelength ranges between 2,000 and 2,900Å; while
in the extreme ultraviolet (between 2,000 and 1,300Å)
this role is taken over by the ordinary diatomic oxygen
molecules (O_2). These molecules are so voracious de-
vourers of UV-quanta that unit optical depth (signi-
fying an attenuation of the incident flux to 37 per cent
of its original intensity) is reached already at altitudes
close to 327,000 ft., where the residual air density is
only one-millionth of that at sea-level.

In the wavelength band between 1,300 and 1,100Å
the atmospheric absorption fluctuates markedly, and
offers a number of narrow windows of reduced opacity.

One of these windows includes the principal hydrogen line Lyman-α (λ 1,216Å), whose light can penetrate as low as within 262,000 ft. above the surface. At still shorter wavelengths (between 1,000 and 400Å) atmospheric absorption by ionized molecular as well as atomic nitrogen and oxygen becomes so intense that the altitude of the level of unit optical depth increases to some 524,000 ft. (where the air density diminishes to 10^{-9} of that at sea level). For λ < 400Å this absorption becomes less important; and X-rays from cosmic sources can penetrate to altitudes between 229,000 and 262,000 ft. and occasionally still deeper.

Towards the red end of the optical window, atmospheric transparency becomes increasingly impeded by the absorption of the molecules of water vapor and carbon dioxide, which blots out most parts of the spectrum at wavelengths between 1 and 1,000μ's. This absorption does not, to be sure, set in as abruptly as does the ozone absorption in the UV below 2,900Å; for the first water vapor bands appear already below 1μ (i.e. 10,000Å); and at longer wavelengths there are at least two sizeable windows (between 8 and 12μ and around 25μ), where some infrared light from space can actually reach the ground. However, beyond λ ~ 30μ the absorption becomes once more so impenetrable that even the Sun will not begin to emerge through the atmospheric curtain till at sub-millimeter wavelengths.

Between 1 mm. and a few meters—in the domain of the radio-spectrum—the transparency becomes wellnigh complete, until it is eventually cut off by the reflecting properties of the ionosphere. Charged particles of the same ionized layer, which acts as a concave mirror reflecting back to Earth short-wave radio signals of ground-based transmitters, will act as a convex sphere in reflecting back similar radio waves reaching our Earth from space; to observe these we would have to lift our receivers above the altitudes of the E and F layers.

To summarize—apart from the relatively narrow "optical window" between 3,000 and 10,000Å, and a much wider "radio window" between 1 and 10,000 mm., our atmosphere on ground is almost totally opaque to incident light reaching us from outer space; and in order to break through these barriers it is not sufficient to ascend merely above the main atmospheric air mass, but attain altitudes at which the air density becomes less than 10^{-8} or 10^{-9} of that at sea-level.

In order to appreciate the full seriousness of this selective absorption, attention is invited to the graph on page 42, which shows the distribution of energy in the spectra of a star akin to that of the Sun (GO-V), of surface temperature close to 6,000°, and of an early-type Main Sequence star (of spectrum BO-V) of surface temperature about 30,000°. For the former, most part of radiant energy emitted from the surface falls within the domain of the atmospheric "optical window" (shaded on our diagram); while for the latter most light is emitted at frequencies too high to penetrate the atmospheric ozone, and is thus completely concealed from view of

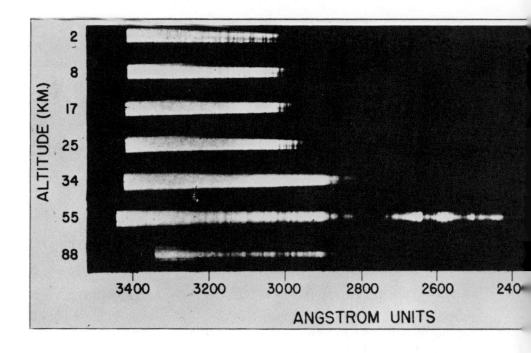

a ground-based observer. Most "hot" objects in the sky (of particular interest to the astrophysicist) are similarly affected—this is why our observational acquaintance with them has so far been very incomplete.

On the other hand, the planetologists of the solar system—concerned as they are with the radiation of "cold" bodies with temperatures mostly between 100° and 400°K—are similarly hampered by our limited access to infrared spectra of such objects (where most part

A

B

C

D

E

F

G

2200

Historic rocket spectra of the Sun obtained on October 10, 1946, by Tousey and his collaborators, at different atmospheric altitudes.

of their energy will be emitted), due again to the absorption of atmospheric water vapor and carbon dioxide. In brief, the combined effect of atmospheric optics is to deplete the light passing through it of some of its most informative parts to such an extent that—like Icarus—astronomers have long dreamed of spreading their wings into space to lift their instruments above the atmosphere; but unlike Icarus, they acquired of late better technical means for accomplishing their purpose.

The aim of the chapters which follow will be not only to review briefly what has been done to this end so far, but also to share with the reader our expectations and plans for the future.

Of the two barriers defining the "optical window" the one at the red end is easier to surmount; for absorption of water vapor (and, to a lesser extent, carbon dioxide) effectively disappears at altitudes between 82,000 and 128,000 ft.—i.e. below the lower limit at which the ozone absorption sets in. Therefore, in the infrared space begins at generally less than 98,000 ft. above sea-level. However, to observe celestial bodies in the entire ultra-violet and soft X-ray domain of the spectrum would be possible from a vantage-point elevated no less than some 524,000 ft. above the surface; and almost the same altitudes would be required to open up the longwave end of the "radio window".

It is intriguing (though idle) to speculate how much better off we—the astronomers—would be if oxygen and its compounds were totally absent from the air. In such a case, the continuous region of atmospheric transparency would extend from wavelengths of about 1,000Å to some 10 m.—i.e. ranging by a factor 10^8 or over nine octaves of the spectrum of electromagnetic radiation; only the extreme ends below 1,000Å and above 10 m. would continue to be blotted out by the absorption of ionized nitrogen, and by the scattering on free electrons liberated by this ionization. This is, however, we repeat, an idle speculation; for without atmospheric oxygen (which may, indeed, be all of organic origin) no comprehending eye could contemplate the grand panorama of the Universe around us with the curtains fully drawn up; all the precious rare quanta would be wasted on dead rocks.

To summarize, the loss caused by absorption of light from celestial sources passing through the atmosphere is extensive and irreparable; and most of it occurs very high in the atmosphere—well above the main air mass. Moreover, the absorption itself does not by any means constitute the only (or even the principal) source of damage to the telescopic images observable on ground; for much more serious damage arises from the refraction and diffraction of light passing through tropospheric layers of increasing air density, a few miles above the terrestrial surface.

Consider the fortunes of a light ray (of a wavelength which can penetrate to the ground) from a celestial body on the last leg of its long journey through space, in a few dozen microseconds before it will be trapped by the telescope. During this time it has got to negotiate its way through air of increasing density which is in a perpetual motion, and to be repeatedly deflected from its course by the local fluctuations in the "index of refraction." This latter quantity plays a crucial role in any attempt to trace the path of the incident ray, or the corrugation of the advancing wave-front. This refraction index depends slightly but sensitively on the local density and temperature, as well as on the wavelength. Now, the

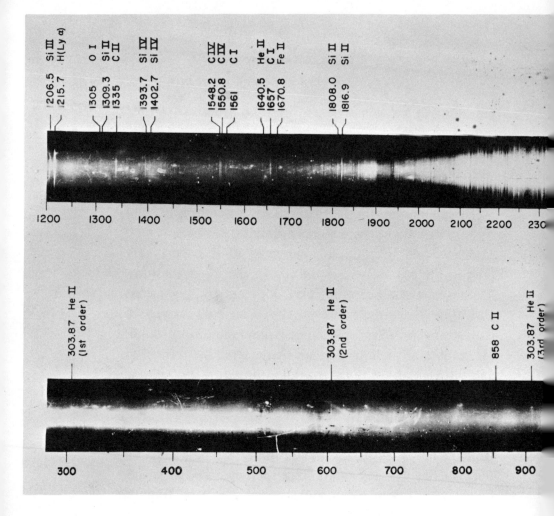

atmospheric density and temperature not only vary sys-
tematically with the height, but also laterally, due to
motions which agitate the air and are essentially turbulent to a very great height—as attested by wellnigh com-

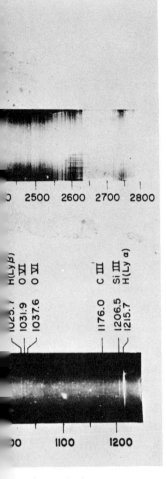

Solar spectrum from 300Å to 2800 Å recorded at an altitude of approximately 136 miles above New Mexico on June 4, 1958 by the University of Colorado. (By courtesy of William Rense).

plete mixing of atmospheric constituents of very different molecular weight.

Of fluctuations of the atmospheric index of refraction due to turbulent motion, the astronomically most insidious are those caused by local turbulence on a scale comparable with, or smaller than, the aperture of the telescope. The ripples in a turbulent layer of air some few miles above us constantly form and dissolve on a very short time-scale; and their combined action as transient cylindrical lenses give rise to the familiar phenomenon of "scintillation".

The reader is doubtless familiar from his own experience with the way in which the stellar images shimmer on a clear night, especially near the horizon; and more quantitative measurements of this phenomenon have revealed a number of interesting facts. First, the stellar images as seen through the atmosphere exhibit rapid changes of intensity (brightness scintillation) of an amplitude which increases rapidly with diminishing aperture (or, for a fixed aperture, with diminishing angular diameter of the light source). It is a well-known fact that a planet in the sky can be easily distinguished from the stars with the naked eye because the latter scintillate, while the former shine with a light that is much more steady; and considering the known sizes of their apparent discs, we can identify the disturbing air elements with turbulent cells that are only inches in size, and a few miles overhead. The brightest scintillation appears to be due mostly to diffraction of light on such air blobs; and the fact that stars scintillate also in color reveals that scintillations in different frequencies are not coherent.

On the other hand, anomalous refraction of light in the same kind of short-lived air blobs produce lateral as well as vertical movements of the actual image of the star in the neighborhood of the mean focus, giving rise to the so-called "shimmer discs". In order to illustrate what we obtain if we try to record such a dancing image with the detector in a fixed position near the focus, let us look at page 36 which features a sequence of four images of a first-magnitude star (which happens to be Aldebaran, the brightest star in the constellation of Taurus), televised in rapid succession by Kuprevich at the Pulkovo Observatory with the aid of an optical system of equivalent focal length of 125 m. The exposures were one-fiftieth of a second, and the individual frames were taken 10 sec. apart. A glance at the irregular anatomy of the fluctuating images reveals better than any verbal description the damage inflicted by the vagaries of atmospheric refraction and diffraction on the image of a light point!

To give another example of atmospheric distortion of stellar images under even the best possible conditions, page 37 reproduces a series of photographic records of the same star, taken by Rösch with the Lallemand electronic camera in the focus of the 43-inch reflector of the Observatoire du Pic-du-Midi in the Pyrenées (see page 183; individual exposures of 2 millisec. were repeated at a time interval of 25 millisec. In each case, an irregular lump of silver grains represents an imprint of what above the atmosphere would appear as a light point but a fraction of a millimeter in size on the scale of our reproduction. Such is the loss of resolution brought about by fluctuating atmospheric refraction anomalies under which we, the ground-based plodders have to labor, dreaming of space where starlight does not dance. "Twinkle, twinkle, little star" may be an amusing ditty to the child; but one often capable of driving an astronomer to distraction.

This ceaseless agitation of celestial images, due to fluctuating refraction and diffraction of air masses above the telescope, represents the most serious limit to astronomical work from ground-based facilities. It never completely subsides; but its frequency-spectrum may vary from place to place, or from time to time. When the "seeing" is termed to be bad, high-frequency components (10-100 Hz.) predominate; and at such times very short exposures are required to stop it—even though all we may accomplish thereby is to freeze a distorted image. At first-class observing sites—such as the Observatoire du Pic-du-Midi (see page 183)—the average frequency of atmospheric oscillations diminishes to 0.1 Hz.; and it is under such conditions (which can be monitored by human eye) that images of best quality are often obtained.

In order to place the role of this "atmospheric seeing" in proper perspective let us recall that no telescope of finite size could image a star as a light point of proper angular diameter. In accordance with the wave nature of light, an optically perfect objective or mirror of aperture D will image a light point as a disc of an angular diameter equal (very nearly) to $1.22''$ (λ/D), where λ

After sunset on the Moon—the streamers of the solar corona as recorded by Surveyor I above the lunar horizon in June 1966. (By courtesy of JPL, CIT.)

denotes the effective wavelength of observation. This disc should, for perfect optics, contain about 84 per cent of the intensity of the original light point; the remainder being confined to a system of discrete concentric fringes ("Airy's rings") surrounding the central core. With deteriorating quality of the optics the high concentration of light in the central disc will, however, be lessened and smeared out, together with the rings, into a much larger configuration.

Accordingly, a 24-inch aperture should, for λ 5,600Å (yellow light), image a light point as a central disc 0.24" across. A 40-inch aperture would reduce this apparent diameter to 0.15"; while a perfect 200-inch mirror should be theoretically capable of imaging discs to 0.03" in size. In actual fact, however, the ever-present atmospheric disturbances seldom permit ground-based telescopes with apertures in excess of 16-20 inches to attain the angular resolution permissible by the geometrical optics; and the actual resolution of the 200-inch telescope at Palomar Mountain is seldom less than 0.5"—due to partly optical, but mainly atmospheric reasons.

For the first two hundred years of telescopic astronomy atmospheric degradation of images was but of secondary concern to the observers wrestling with the optical and mechanical imperfections of their instruments. Technology marched on, however; and the stage at which the "atmospheric seeing" became the limiting factor of astronomical observations was reached some time around the turn of this century. It set off an exodus of the observatories away from large cities—a task made all the more imperative by the gradual degradation of their environment and increasing ease of transport—and in quest of locations blessed with a better steadiness of seeing and sufficiently high percentage of clear skies.

This exodus, which at present has almost been completed, led astronomers to many parts of the world; and there were powerful reasons behind it. The modern astronomical telescope is an expensive instrument; and rapid advances of technology are tending to shorten its useful lifetime. When the requisite capital investment, operating costs as well as depreciation, are taken into account, a single observing night with a telescope of an aperture around 100 inches or more can cost several thousand dollars. To erect such a telescope in a locality where one can expect (say) twenty clear nights per year would render each hour of observation ten times as expensive as in a climate where two hundred such nights can be expected; and, similarly, the light lost through poor seeing on the jaws of the slit of a high-dispersion spectrograph can become quite an expensive commodity.

Guided by such considerations, astronomers of the present generation concluded regretfully that there are no really suitable places for large telescopes in Europe north of the Alps; nor are there in North America east of the Rockies. Southern Europe, Africa, South America, Australia—all offer possibilities which are being intensively investigated at the present time; and the majority of large ground-based telescopes of the future are likely

to be located in those areas.

A judicious choice of the observing site can increase the number of night-hours when it may be possible to use the telescope; and superior quality of the images controlled by local "boundary conditions" can increase its efficiency. However, it has become clear to astronomers in the past fifty years and more that the right direction in which their telescopes should be deployed is upwards—as high above the sea-level as is permitted by logistic considerations.

The Lick Observatory at Mount Hamilton (4,209 ft.) in California, with its 36-inch refractor in 1878, became the forerunner of many subsequent observatories (Mount

The crab Nebula in the constellation of Taurus, remnant of the Chinese supernova of the year 1054 AD, photographed in the light of the hydrogen line Ha with the 200-inch telescope of the Palomar Mountain Observatory.

Wilson, 1908, etc.) built at moderate altitudes above the surrounding landscape. In Europe the classical observatory of this type is located at Pic-du-Midi in the French Pyrenées, 9,387 ft. above sea-level; but even at the highest active observatory in the world—at Sphinx, Jungfraujoch, in the Swiss Alps—at 11,729 ft. the air pressure still remains not less than two-thirds of that at sea-level. Temporarily, astronomical observations have been made from higher sites—thus in the early years of this century Jules Janssen maintained for a time an observing station on the top of Mont Blanc (15,781 ft.); but its site became eventually untenable by the motion of the glaciers. More recently, important observations requiring high atmospheric transparency were repeatedly made at Chacaltaya (altitude 18,007 ft.) in the Bolivian Andes; but no permanent instruments have so far been erected at that site; and while even there 53 per cent of the atmospheric air mass still remains overhead, the width of its "optical window" has scarcely been altered.

In an effort to free ourselves from the limitations of the atmosphere, ascent of any terrestrial mountains with our instruments remains a mere makeshift as far as our main objectives are concerned. In order to get above the turbulent air—let alone to widen the windows of atmospheric transparency—we must take our instruments much higher into space—without the possibility of solid support; and in the next chapter we shall give a brief account of how far we have reached already in this direction in our endless quest.

3 · On the Threshold of Space: Balloons

Many reasons outlined in the preceding chapter make it obvious why astronomers should have been among the first Earth-born scientists to feel an irresistible urge to spread their wings into space. Nevertheless, respect for historical truth leads us to confess that this was not exactly the case. What is worse, an impartial inquiry would be bound to reveal that, when it comes to utilizing to their ends the new means which the continuing advances of technology place from time to time to the astronomer's hands, the astronomers have, on the whole, been a pretty sedentary lot. Not so the theoreticians, perhaps, who by their thought have measured the depths of space long before it could be followed up by any instruments; but the observers have traditionally preferred to play it safe.

The examples in support of this thesis are indeed too many to be quoted here in full. Think only of Johannes Kepler, who—far ahead of the space age—travelled to the Moon on the wings of his demons through the shadow-cone connecting us with our satellite during lunar eclipses; while, at the same time, Galileo Galilei had a hard time making his contemporary peripatetic colleagues even look through his telescope; the latter being fearful lest what they would see might deprive them of the comfort of holding on to their Aristotelian preconceptions. It took the better half of the seventeenth century before astronomers had acquiesced in this new device (though, for positional measurements, Hevelius preferred the naked eye well after 1650); and the notion that "astronomy is what you see through a telescope" was indeed slow to be born, as it is slow to fade out from the minds of our more conservative confrères today.

William Herschel's reflectors from the turn of the eighteenth century made little impact on the astronomy of subsequent generations till almost a hundred years later. And even within our own lifetime the art of mounting and controlling large telescopes lagged far behind the "state of the art" of contemporary technology; and until about 1940 the principal servo-mechanism for telescope control was still the (ill-paid) human hand. The 200-inch glass giant of Palomar Mountain was the first large telescope in which the optical perfection was matched by that of its mounting and control; and as much as any other telescope marks the epoch by which automation entered the domain of observational astronomy.

In view of these facts reflecting the ways and prejudices of our professional life, it is perhaps not surprising that astronomers were not in the front-line of space research from the beginning, waiting at the head of the line for a ride for themselves or their instruments. This place of honor belongs, instead, to geophysicists concerned at first mainly with the studies of the upper atmosphere; and to nuclear physicists whose quest of cosmic rays led them through increasing altitudes above the main atmospheric air mass to the threshold of space. It was these tasks which brought up a whole generation of physicists conversant with instrumentation appropriate for space vehicles—a task with which astronomers are still catching up at the present time.

Among the first vehicles which offered transport to scientific instruments above the troposphere, the primacy belongs to the *balloons*. To give even a briefest outline of the history of the lighter-than-air vehicles to reach high altitudes is wholly beyond the scope of this chapter; for we are really concerned with such devices only in so far as they carried telescopes for astronomical observations. For lack of space—though not of admiration—we thus have to forego with briefest mention the historic flights of the stratospheric balloons of Auguste Piccard, in which this intrepid investigator and his colleagues ascended around 1930 to altitudes close to 70,000 ft. in pursuit of the primary cosmic rays—altitudes exceeded in 1935 by the American Air Force officers Stevens and Anderson, who on November 11 of that year piloted their balloon (Explorer II) to an altitude of 72,400 ft. to set a record that held for many years to come.

In postwar years substantially higher altitudes were attained by unmanned balloons launched mainly for meteorological purposes, the ceiling of which (for nominal payloads) has been increased to some 140,000 ft. However, by the mid-1950s astronomers were no longer far behind the meteorologists; and their principal target became at first the Sun. That this should be so was only logical; for the principal curse of ground-based observation—atmospheric turbulence—is aggravated in the immediate proximity of the brilliant disc of the Sun, due to its heating effect. And yet, at the moments of high resolution, the solar disc has revealed an intriguing and transient microstructure of great interest to the students of solar physics. At the same time the high intensity of its light—permitting very short exposures—minimized demands on stabilization and control of the telescopic space platform; for even the blindest "photoelectric eye" of a simple servo-mechanism could localize its target and help to point it in the right direction. No wonder that the Sun has become an early object of attention for balloon-borne telescopes!

The first such attempts in postwar years were made in Europe, where, in November 1956 and later in April 1957, D. E. Blackwell and D. W. Dewhirst of the Cambridge University Observatory, together with A. Dollfus of Meudon Observatory in Paris, photographed successfully the granulation of the solar surface, altitudes between 20,000 and 25,000 ft., with the aid of an 11-inch refractor. Theirs was a manned balloon; and the need to provide for life support imposed a relatively low ceiling for attainable maximum altitude.

The latter was, however, to be drastically relaxed in the impending American project (Stratoscope I), undertaken at about the same time by a Princeton University group under the direction of Martin Schwarzschild. In order to attain substantially greater altitudes, an unmanned balloon was used as the principal carrier: namely, a U. S. Navy Skyhook (of some 10^6 cu. ft. volume), capable of lifting a 1,400 lb. payload to an altitude up to 84,000 ft. The central part of this payload was a 12-inch mirror telescope of f/8 focal ratio, which by suitable optical enlargement yielded an image corresponding to an effective focal length of 200 ft. The telescope itself was only 9½ ft. long and weighed 300 lb. The size of the solar image formed in the effective focus was approximately 2 ft. across; and only a small part of it could be recorded on any frame of its automatically operated 35 mm. film camera. In pointing the telescope, several pairs of photo-diodes were employed to act in concert as remote "eyes" to find the Sun, and center any part of its disc on the film in both azimuth and elevation. The intensive heating of the system by concentrated sunlight precluded continuous taking of the exposures; and the position of the focus itself had to be established essentially by trial and error.

Within these limitations, this project proved technically to be an almost unqualified success. Altogether five flights of Stratoscope I were undertaken—three in 1957 (on August 22, September 25 and October 17) and two in 1959 (on July 11 and August 17)—of which especially the last one furnished excellent photographs of the sunspots and solar granulation, taken from an altitude of close to 80,000 ft.

The success of this first Princeton balloon experiment encouraged other groups of investigators to embark on other aspects of solar studies which the atmosphere above us makes extremely difficult from the ground: namely, to explore from balloon altitudes the outer layers of the Sun and its immediate surroundings. As is well known, the mass of the Sun does not really end up abruptly at any particular distance from its center, but rather peters out through its corona into the zodiacal cloud. What we usually call the "solar surface" is merely a layer at which the solar gases cease to be opaque in visible light; though if our eyes were sensitive to the radio-frequencies, for instance, the Sun in the sky would appear to be several times as large.

The principal reason why this extension of the Sun is so difficult to observe from the ground is, of course, the fact that its relatively low surface brightness is so very largely drowned in the light of the brilliant Sun disc scattered on the molecules of our atmosphere. The total amount of visible light emitted by the solar corona constitutes only a few parts in a million of total sunlight (i.e. comparable with that of full Moon rather than the daylight Sun); and very exceptional atmospheric (as well as instrumental) conditions are needed to catch any glimpse of it from the ground outside the rare and fleeting moments of total eclipses.

It was the quest for such conditions that led Bernard Lyot to the top of Pic-du-Midi in the 1930s; and it was the same quest which led more recently G. Newkirk and J. D. Bohlin, astronomers from the High Altitude Observatory at Boulder, Colorado, to organize the launch of special balloons (the Coronascopes) to photograph the solar corona in full daylight. The first such attempt (Coronascope I) was made in the autumn of 1960, when balloons were made twice (on September 10 and October 3) to carry a small coronograph (of 1.3 in. free aperture and f/10 effective focal ratio) to an altitude of 80,000 ft. and more. The images were recorded automatically on a 35 mm. red-sensitive 1N Kodak Spectroscopic film (with maximum sensitivity around 8,300Å) to minimize further the effects of light scattering on air molecules (which by Rayleigh's theory is known to diminish with inverse fourth power of the wavelength).

The first flights of the Coronascope I in the autumn of 1960 attained an altitude close to 80,000 ft.; and a repetition of the experiment on March 5, 1964 with improved equipment, reaching a ceiling of 99,000 ft., met with full success; for it recorded—for the first time outside of total eclipse—coronal streamers up to a distance between 2-5 solar radii from its center. The success was primarily due to the high altitude to which the coronograph was lifted by the balloon; for the amount of the parasitic light scattering is proportional to the residual air mass overhead; and the latter diminishes exponentially with increasing height. Thus at an altitude of 20,000 ft. above sea-level, some 47 per cent of air still remains overhead; at 40,000 ft., this amount has been diminished to 20 per cent; and to 3 per cent at 80,000 ft. From this it follows, however, at 100,000 ft.

(i.e. near the ceiling of Coronascope II flight on March 5, 1964) the sky brightness in the red was still two to three times as large as at sea-level during total eclipses of the Sun—in order to match the eclipse conditions we should have to go still higher! (see page 41.)

Let us add quickly that this has already been done—in fact, before the flight of Coronascope II—when F. C. Gillett, W. A. Stein and E. P. Ney of the University of Minnesota, working in collaboration with V. D. Hopper and J. Sparrow of the University of Melbourne, flew on July 20, 1963 several balloon-borne fast cameras with focal lengths of 23 mm. (f/0.6), 35 mm. (f/2), and 189 mm. (f/6) to an altitude of 110,000 ft. to study the far extension of the Sun and the zodiacal light; and the absolute intensity of the light of the solar corona between the limb of the Sun and 70° elongation was determined from such photographs.

The cameras with which all this work was performed were, however, but small fry in comparison with another concurrent project in Princeton. Following the earlier success of the special-purpose Stratoscope I, Schwarz-schild and his colleagues—in collaboration with the Perkin-Elmer Company in Norwalk, Conn.—set out to design and build a multi-purpose astronomical balloon facility under the code name of "Stratoscope II". A larger balloon vehicle was used to lift to the ceiling altitude of some 85,000 ft. an increased payload of 6,300 lb., which comprised a 36-inch f/4 fused silica mirror of excellent optical qualities, as part of a modified Grego-

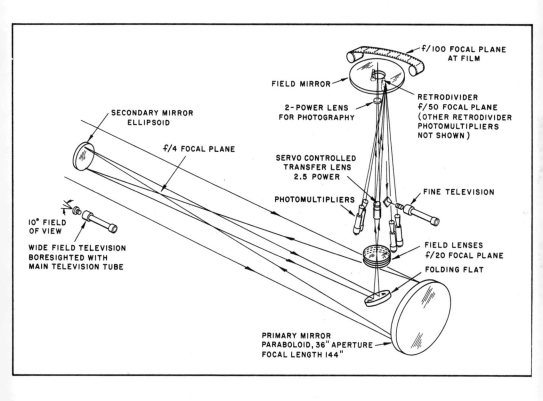

Stratoscope II optical system.

rian telescope of an equivalent focal length of 300 ft.; and an elaborate television system was incorporated to guide and control the telescope. (see page 131.)

The first two flights of this elaborate instrument devoted to infrared spectroscopy (with the air of an airborne germanium bolometer) took place in 1963. The first (on March 1), encountered some severe technical difficulties; but a low-dispersion spectrograph recorded a part of the IR spectrum of Mars, an analysis of which enabled the investigators to refine an upper limit on possible water-vapor contents of its atmosphere: according to this result, Martian air (which is essentially pure carbon dioxide) cannot contain more than 0.1 per cent of the terrestrial humidity. The second flight of Stratoscope II—the most successful one so far—took place on November 26, 1963; and during several hours of a cruise at altitudes ranging between 71,000 and 84,000 ft., the near infrared spectra (between 1 and 3μ) were scanned for nine red giant stars (which included Aldebaran and Betelgeuze); apart from some additional scans of Jupiter and of the Moon.

The next three flights of Stratoscope II in 1964, 1965, and 1966 failed to attain their objectives because of various mechanical problems; and it was not till with the sixth flight on May 18–19, 1968, that the fortunes of the project met with better success. A direct photograph of an external galaxy (NGC 4151), obtained in the course of this flight, attained an angular resolution of 0.3″—no better, perhaps, than that obtainable from the ground under good seeing conditions (and about three times worse than the mirror of Stratoscope II should have theoretically attained), but the performance of the automatic guiding of the plate during exposure (better than within 0.05″) exceeded expectations, and augurs well for the future.

Another heroic episode in the story of contemporary balloon astronomy has been written in connection with current efforts to determine by spectroscopic observations the contents of water vapor in the atmosphere of Venus. The difficulty of this task rests on the fact that, in seeking to establish the presence of the absorption bands of water vapor in the spectrum of another planet, we have to observe it through our own atmosphere, which contains the same compound in ample measure—much greater than Venus can possibly possess above the perpetual veil of its clouds. With the carbon dioxide it is the other way around; and its spectroscopic identification (in large amount) in the Cytherean atmosphere by Adams and Dunham goes back to 1932. But if the amount of H_2O in the Cytherean air is small, how to establish a superposition of its effects on the background of more powerful telluric bands?

The only way to do so is to take advantage of the fact that, on account of the relative motion of the Earth and of Venus, the Cytherean spectral lines should be generally displaced with respect to the telluric lines by Doppler shifts corresponding to radial velocity differences oscillating between ± 35 km./sec. In the near infrared (where absorption of water molecules is concentrated) this velocity difference should in the red produce Doppler shifts within $\pm0.9\text{Å}$—which would be easily measurable if the telluric bands were not so strong. As it is, two avenues of approach remain open to us: either to concentrate on high-dispersion studies of band structure in the part of the spectrum which remains accessible from the ground (where alone a sufficiently high dispersion can be employed so far to permit a detection of line splitting), or—if scarcity of the material makes resort to strong (saturated) lines necessary—to make observations from sufficient altitude to leave most of the terrestrial absorbing material below us; for only in such a case is there a chance to establish the Cytherean component from the asymmetry of a composite profile due to a superposition of Doppler-shifted telluric and Cytherean lines.

Both these methods were followed in recent years in repeated efforts to detect moisture in the Cytherean atmosphere; though the second method interests us here in the first place. It is this method which led the octogenarian Jules Janssen to the top of Mont Blanc with his young assistant Stefánik and Millochau in the first decade of this century to distinguish between telluric and solar absorption lines from the different intensities of the former at Meudon and 15,777 ft. above sea-level; and the same quest, half a century later led A. Dollfus in France (1959) and John Strong in the United States (1964) to undertake balloon flights in quest of water vapor on Venus by observations of the saturated water bands at $\lambda\lambda 1.13$ and 1.4μ from much greater altitudes.

On November 29, 1959 the U. S. Navy sent a manned balloon (piloted by Commander Malcolm D. Ross, with Charles B. Moore acting as scientific observer) to an altitude of 80,000 ft. to search for the asymmetry of the 1.13μ water band; and from the results of this experiment Strong and Benedict estimated the amount of water vapor above the Cytherean cloud layer to be equivalent to about 20 microns of precipitable water per square centimeter—comparable with the amount found above the high-level stratospheric clouds on the Earth. In 1964 Strong and his group repeated the same experiment by flying twice (on February 21 and October 28) unmanned balloons carrying a 12-inch telescope to an altitude close to 87,000 ft., with the results comparable with their earlier findings.

It may be added that, in the summer of 1966, two groups of Earth-bound observers (Belton with Hunten from the Kitt Peak National Observatory, and Spinrad with Shawl from Berkeley) have at last got a glimpse of Cytherean water vapor from the observed asymmetry of the $\lambda\lambda 8,189$ and $8,193$Å lines that are visible on the ground, amounting to about 60 microns of precipitable water (Spinrad). These abundances are still preliminary; and their actual amount may also vary with the (Cytherean) season. But, unlike on Mars (where the Stratoscope II on March 1, 1963 failed to detect any atmospheric humidity exceeding a few microns of pre-

cipitable water), manned as well as unmanned balloon
flights in the last few years established the presence of
ten to twenty times as much of it above the clouds on
Venus—a result corroborated by subsequent ground-
based work.

Perhaps the most dramatic contribution of the balloons
to space-age astronomy has, however, been the recent
investigation of an X-ray source in the Crab Nebula
(see page 58). The latter—a remnant of the Chinese
supernova of 1054 AD—represents one of the most enigmatic as well as instructive experiments performed by
Nature in the recent past at our veritable doorstep in
space (only about 1,000 parsecs or 3,300 light-years
away—so that the actual outburst of light which reached
us in 1054 AD actually occurred some time during the
twenty-third century BC); for apparently a whole star
committed then a suicide which led to a total dispersal
of its mass into space. Nothing has advanced our knowledge of the last throes of a star's life—and nothing
brought astronomy and physics closer together in the
last twenty years—than a postmortem inquest on the
Crab Nebula which is still going on; and the astronomical
evidence on the radiation emitted by this enigmatic object
spans a spectral range extending from 1 m. radio-waves
to X-rays of less than 1Å wavelength (i.e. over a range
of more than 10^{10} in λ).

Most of this evidence has been gathered so far, to be
sure, through intermittent gaps of atmospheric transparency from the ground; and as such it does not belong

legitimately within the scope of this book. A discovery of an intense source of X-rays within the Crab Nebula was, however, not possible till our instruments could be lifted above the main atmospheric air mass; and although the first glimpse of this source was obtained from rockets (in June 1963, by Friedman and his associates from the U. S. Naval Research Laboratory)—about which more will be said in the next chapter—it could since be observed also from balloon altitudes; and as this feat represents a veritable *tour de force* of balloon astronomy so far, this experiment deserves more than a passing mention in this place.

In a balloon launched by the M.I.T. group of investigators (Clark and his colleagues) on July 21, 1964 to an altitude of over 130,000 ft., X-rays with energies between 15 and 60 keV (i.e. $\lambda\lambda$0.2-0.8Å) were observed to emanate from a source coinciding in position with the Crab Nebula; and the intensity of the radiation was measured quantitatively by means of scintillation counters. We shall return to the significance of these measurements in the next chapter, in connection with the parallel investigations carried out by means of the rockets. For the present, we wish to stress that these experiments, having extended the ceiling of balloon astronomy with appreciable penetration into space at wavelengths below 1Å; and this feat (coupled with a subsequent flight of M.I.T. investigators on February 20, 1965 to observe another celestial X-ray source in Scorpius) have given balloon astronomy its best claim to be considered as a part of space science.

This claim was reinforced further last year, when Haymes and his collaborators at the Rice University in Houston—using a scintillation-type detector lifted by a balloon to an altitude of 125,000 ft.—detected on June 4, 1967 also a flux of γ-rays to emanate from the Crab Nebula (as well as from the Cygnus X-ray source (X1) on which more will be said in the next chapter), on wavelengths as short as 0.022Å (corresponding to photon energies of the order of 0.5 MeV). The Crab Nebula, whose spectrum has now been recorded (albeit with some gaps) over a frequency range of 10^{12} to 1, appears to produce a total power output of about 7×10^{37} ergs/sec.; but its actual mechanism remains still largely unknown.

Summarizing all that has been presented in this chapter, what can we say about the potentialities of balloon astronomy and its prospects for the future? First, we may note with satisfaction that, by 1968, the capabilities of balloon vehicles have progressed far enough to lift payloads of the order of a few tons to altitudes between 80,000 and 85,000 ft.; up to almost a ton to 100,000 ft.; and some 220 lb. to 130,000 ft. While the stabilization of their observing platforms still leaves something to be desired, flights lasting for several hours have been accomplished at these altitudes; and flights above 140,000 or even 150,000 ft. have already been achieved, though at much greater penalty in payload and reliability. Moreover, several current lines of technical developments —both in balloon materials and structural design—give promise of further increase of performance in the future.

What have we gained by this ability to lift our tele-scopes to altitudes which would have looked stupendous only a generation ago? At 100,000 ft. only 1 per cent of the entire atmospheric air mass still remains over-head; but even so the sky is not yet completely black; it would, in fact, appear brighter to us than that seen at sea-level during a total eclipse of the Sun; only brighter stars would be visible to the naked eye on its back-ground. At 80,000 ft. and above all deleterious effects of unsteady "seeing," due to atmospheric micro-turbu-lence and deplored in the last chapter, are left far below our cruising altitude—not that all turbulence would have died down (the air remains turbulent to much greater altitudes, as is demonstrated by continued well-nigh perfect mixing of its various constituents of very different molecular weight), but because the air still overhead is too rarefied to distort the oncoming light waves through anomalous refraction to any appreciable extent. In other words, stars would not "scintillate" any more at balloon altitudes; and the telescopic images in the focal plane would depend essentially on the quality of the optics. This is why the Princeton investigators insisted on diffraction-limited mirrors for their Strato-scopes; and why no other mirrors would do under con-ditions where perfect "seeing"—so rare on the ground—becomes standard.

However, when we come to consider the transparency of the residual air mass above the altitudes attainable by

A photograph (taken around 1880)
of Lord Rosse's 72-inch telescope
of 53 ft. focal length (f/9), erected
at Birr Castle, Ireland, in 1845.

ballons, it appears that the remaining few per cent of
atmosphere do not relax much their hold on the width of

the "optical window" through which we can view the Universe. An ascent to 100,000 ft. clears up largely the water-vapor absorption in the near infrared (at wavelengths between 1 and 5μ), and weakens the CO_2 absorption bands between 5 and 8μ; but the violet edge of the optical window has been barely affected (for the balloons only skirt the lower limit of atmospheric ozone). It is the opening up of the X-ray window (below 1Å) which promises to give balloon astronomy its principal lease on life.

Another advantage of astronomical balloon work is the fact that its results (and even the apparatus) are largely directly recoverable, and less dependent on automatic scanning and telemetry. It should, however, be also kept in mind that although perfect seeing should invariably prevail at balloon altitudes, the atmospheric conditions on ground suitable for the launching of large balloons are less frequent. In fact—judging from the record—they may be almost as rare as the conditions required for good-quality images at first-class astronomical sites on the ground. The photographs of solar granulation taken aboard Stratoscope I—admirable as they are—did not reveal any new phenomena not previously observed from the ground. In fact, photographs virtually as good were previously secured from ground-based observatories (in particular, from Pic-du-Midi), though only on very rare occasions.

The 40-inch refractor at the Yerkes Observatory of the University of Chicago—the largest astronomical refractor of the world.

The costs of launching a large balloon with recoverable payload, while relatively large in comparison with those of ground-based telescopes (the Stratoscope II project alone has entailed an expenditure of over 2½ million dollars so far), are small in comparison with those of the launch of artificial satellites and orbiting observatories (of which more will be said in the next chapter). But if one considers the fact that a close Earth satellite operative for one year can provide us with some 4,300 hours of net observing time, there is no doubt that a unit of observing time will cost less aboard a satellite than aboard a balloon. Above all, the fact that balloon work still remains largely constrained by the same atmospheric absorption limits as we experience on the ground renders it of rather marginal significance for the future of space-age astronomy.

The usefulness of balloons to serve as pilots for deep-space experiments remains uncontested; so is their advantage of a relatively short lead-time, enabling new experiments to be planned and carried out in quick response to new discoveries and technical developments. Even so, however, all this renders the balloons of ancillary rather than prime significance for future progress of our science. Balloons can lift our instruments to the threshold of space; but being lighter than air, scarcely beyond. Further penetration into space will have to depend on vehicles whose lifting power is based on principles quite different from buoyancy; and to these we wish now to turn our attention.

4 · Above the Ionosphere: Rockets and Satellites

In conclusion of the preceding chapter, we emphasized that balloons have only a partial right to be regarded as space vehicles; for of the limitations imposed on ground-based astronomy by the atmosphere above our heads only one can be satisfactorily overcome with their aid: namely, the imperfections of the quality of the images caused by atmospheric refraction anomalies. The ceiling of the altitudes attainable by balloons is sufficient indeed to defeat atmospheric "seeing". However, it is far from sufficient to widen much the atmospheric transmission window in the optical domain of the spectrum; for apart from some relaxation of limits in the near infrared, the picture of the Universe as seen from the balloon altitudes is still very much the same as we see it from the ground. In order to overcome the air absorption more decisively, much higher altitudes must be attained; and this cannot be done by means of the vehicles which are "lighter than air". Substantially higher altitudes can be attained only by vehicles whose lifting power is independent of the atmosphere (in fact, for which the atmosphere itself constitutes an obstacle)—namely, by *rockets*. It will be the aim of this chapter to survey briefly what has been accomplished so far with their aid in the domain of telescopic astronomy.

This is not the place to trace even briefly the birth of
the rocket as the means of propulsion. The roots of the
idea were—as always—manifold; and even more than
the telescope itself the rocket was born as an instrument
of human warfare. First steps in the development of the
rocket probably took place some seven centuries ago,
with the initial utilization of gunpowder-propelled
rockets as "fire arrows" by the Chinese. The latter are
reported to have used them for war purposes as early
as the thirteenth century. As the use of gunpowder spread
throughout Europe in the Middle Ages, so did the em-
ployment of rockets. In the last years of the eighteenth
century the Indian troops used them in their fight against
the British; while the British used them in 1807 to burn
down the city of Copenhagen.

With the improvement of guns and rifles through the
nineteenth century, rockets were gradually abandoned;
and by the time of World War I they had become mili-
tarily obsolete. However, the potentialities inherent in
their use were revived by the work of three independent

*High-altitude observatory at Sphinx,
Jungfraujoch (11,729 ft.) in the Swiss
Alps.*

pioneers: Konstantin E. Tsiolkovsky and his school in Russia in the early part of this century, followed twenty years later by Hermann Oberth in Germany and Robert H. Goddard in the United States. All three envisaged multi-stage rockets using solid as well as liquid fuels as propellants. However, only one of these pioneer lines of work blossomed out to larger dimensions: namely, in Germany, where in 1932 further development of rocket work was taken out of the hands of a small band of enthusiastic pioneers by the military concerned with the procurement of rocket weapons. This was indeed eventually accomplished (as the inhabitants of Great Britain learned to their cost in the last years of the Second World War); and the first successful launch of a V-2 rocket in October 1942 ushered a new era which marks the beginning of our subject.

Following the breakdown of the German armed forces at the end of that war, further development of rocket weapons was brought temporarily to a standstill; but almost a hundred half-finished V-2's captured in western Europe were brought over to the United States, to be put in service for the exploration of the upper atmosphere since 1946; with astronomers close on the heels of the aero-physicists.

The first fully instrumented V-2 rocket was launched successfully on May 10, 1946, from White Sands, N. M. (see Plates 15, 16), to an altitude of about seventy miles to inaugurate the U. S. upper air research program; but its astronomical career began on October 10 of the same year, when the first ultraviolet spectrum of the Sun extending far below the limit of atmospheric ozone absorption was successfully obtained by a group of scientists under R. Tousey of the U. S. Naval Research Laboratory. A small spectograph was mounted in the war head of a V-2 rocket and a number of exposures were automatically made at different altitudes during the flight. On impact, the spectrograph was recovered and its film developed. Although the resolution (of about 3Å) was not as great as subsequently obtained in later flights, spectral features were shown down to wavelengths as short as 2,100Å (see page 46); while all previous solar spectrograms taken from the ground ceased to record any information below 2,900Å.

Since that time extensive solar research has been carried out with the aid of the V-2 rockets and their successors, not only with spectrographs and photographic film, but also with filters and photon counters. In the early days of rocket astronomy, filter techniques were used extensively by Tousey and his associates to single out discrete bands of the solar shortwave spectrum and measure their intensity. To give an example—a beryllium filter of 0.25mm. thickness is completely transparent for soft X-rays ($\lambda \sim 1$Å), but becomes 99 per cent opaque at 6.5Å and 99.9 per cent opaque at 7.5Å. Or an aluminum foil 7½ microns thick possesses transmission properties which are almost identical with the beryllium filter up to 8Å, but for $\lambda > 8$Å its transmission increases abruptly to more than 50 per cent, only to diminish to less than

1 per cent beyond 17Å.

By comparing the photographic density of an image taken through the aluminum foil with that recorded behind a beryllium filter, a comparison of the intensity of solar emission in the region above and below $\lambda = 8$Å can be obtained. At longer wavelengths filters consisting of lithium (or calcium) fluoride LiF (or CaF_2) have been employed to advantage; and the transmitted energy was also recorded on phosphorescent materials (such as layers of calcium or manganese sulphides $CaSO_4$ or $MnSO_4$) which absorb radiant energy for subsequent release by heating. Later on, the phosphorescent detectors were replaced by photon counters (working on a principle similar to that of Geiger counters of corpuscular radiation, and filled with nitric oxide which is photoionized with wavelengths less than 1,300Å), which have been developed by Friedman and his associates since 1950.

In these ways it was established that the intensity of solar radiation, which approximates that of a black body at 6,000° in the visible part of the spectrum, falls below it in the far ultraviolet (in particular, in the domain between 1,100 and 1,600Å), but catches up with it for $\lambda < 1,000$Å, and becomes larger by three orders of magnitude in the domain of the X-rays. Later on, when rockets were stabilized and provided with automatic sunseekers, grazing-incidence spectrographs operated by means of rockets above the ionosphere enabled us to penetrate the solar spectrum with relatively high disper-

The near side of the Moon, showing distribution of the points at which different spacecraft landed in the past nine years.

sion down to wavelengths of approximately 20Å; and many interesting results were obtained, of which a few will be mentioned below.

The hydrogen Lyman-α line (λ 1,216Å) of the solar spectrum was first photographed in this way by Rense in 1953; and the general features of the spectrum (see page 50) were found to be as follows. Between 2,000 and 3,000Å it remains broadly similar to that in the visible, consisting of a continuum of diminishing intensity on which dark absorption lines are superposed. At shorter wavelengths emission lines begin to make their appearance and soon become prominent, while absorption lines gradually disappear; no Fraunhofer lines were observed in the solar spectrum below 1,525Å. In addition to a very strong Lyman-α emission line of hydrogen at 1,216Å, the solar spectrum exhibits many other strong emissions—including the Lyman-α line of ionized helium at 304Å.

As to the intensity of the continuum, the radiation temperature deduced from it proved to diminish to the region of the hydrogen Lyman series, where it appears to be decidedly less than the 6,000° in the visible domain of the spectrum. Below the Lyman limit (at 916Å) it starts, however, to increase rapidly with diminishing wavelength—to attain half a million degrees around 100Å and 1.5 million degrees between 1 and 10Å. These facts disclose that, while the UV radiation from the Sun originates in the chromosphere where temperatures do not differ much yet from the photospheric temperature, solar X-rays are emitted by the solar corona.

Strong evidence in favor of this interpretation was obtained in 1958 by Chubb, Kreplin, Lindsay and Friedman, who launched a series of five rockets during a total eclipse of the Sun on October 12 of that year visible in the South Pacific. Two of these rockets passed through the shadow cone during totality. In the course of this passage their Lyman-α detectors remained virtually inactive, while X-ray detectors recorded only a marginal drop in signal—facts indicating that the source of the X-rays was much less affected by the eclipse—as is indeed true of the solar corona, which is several times as large as the apparent bright disc of the Sun.

In the meantime, rocket propulsion continued to advance with great strides; and the original V-2's were replaced by the second-generation Vikings or Aerobees in the United States, and Skylarks in Europe, capable of reaching higher altitudes, and altitude-controlled by appropriate stabilizing devices to enable at least a rudimentary guiding of the optical instruments in the desired direction. Such rockets had advanced observations of the solar corona out of eclipses in white light (Tousey and his colleagues in 1963) side by side with the parallel balloon work by Newkirk and Eddy (1962), Newkirk and Bohlin (1963) or Ney and his associates (1963, 1964), which we briefly described in the preceding chapter. On June 28, 1963 a small Lyot chronograph was flown by scientists from the U. S. Naval Research Laboratory in an Aerobee rocket to an altitude in excess of

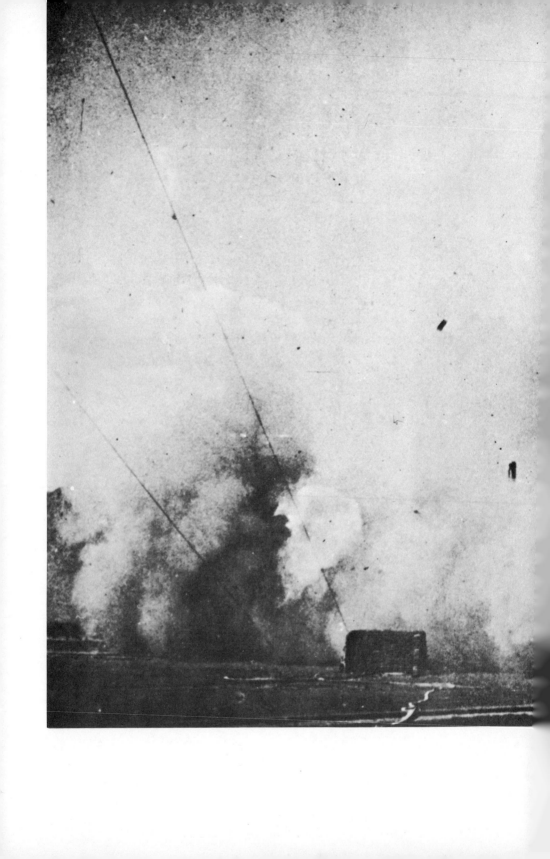

The first V-2 rocket to take to the aid in America from the White Sounds Proving Ground, New Mexico, on 10 May 1946, as a forerunner of greater rockets to come.

655,000 ft.; and twenty-three white-light photographs obtained with exposures between 1 and 56 sec. showed the corona extending up to 10 solar radii from the Sun's center in full daylight. Rocket-born grating spectographs have mapped for us the detailed features of the solar spectrum to a shortwave limit of about 34Å.

However, perhaps the greatest single contribution of rocket astronomy in the past five years has been the discovery of discrete X-ray sources on the Sun (see page 134) and in the sky. If one were to rate scientific discoveries by the novelty of the phenomena which they disclose, the detection of strong X-ray sources outside the solar system has the right to rank as the outstanding achievement of space astronomy to date. Thirty years ago it took radio-astronomy a dozen years to progress from Jansky's original discovery of cosmic noise to the detection of the strongest single radio source Cygnus A. In comparison, since the detection of the first celestial X-ray source in 1962 more than ten additional such objects have already been added to our knowledge; and the fact that none of the celestial phenomena known previously led us to anticipate the existence of X-ray sources approaching even remotely the strength of those observed justifies the expectation that X-ray astronomy of the future will play an important role in advancing our knowledge of the Universe.

How did it come about? As it often happens in the annals of science, almost by a mistake. In June 1962 a group of investigators in the United States, led by Bruno.

Rossi, launched from the U. S. Air Force Proving Ground at White Sands, N. M., an Aerobee rocket instrumented to detect possible X-ray emission (in a band centered around 3Å) from the Moon, caused by impact of the solar wind on the lunar surface.* This the experimenters did not find—and whether or not the Moon actually emits an observable amount of such "brems-strahlung" is still an open question. But, instead—and much to their surprise—they detected a strong X-ray signal from the general direction of the galactic center. Not quite a year later—in April 1963—another group of investigators from the U. S. Naval Research Laboratory, led by Herbert Friedman, launched other Aerobee Rockets carrying X-ray counters ten times as sensitive in the domain of 1-10Å as those employed in the 1962 experiments; and this located a very strong X-ray source in the constellation of Scorpius—the most intense discrete X-ray "star" known so far—and another, though less powerful (about one-eighth as "bright") in the Crab Nebula, of which we spoke already in the last chapter (see page 58).

Since that time more than a dozen other, less intense, sources have been detected on subsequent X-ray scans of the sky; in the course of which approximately 70 per cent of the celestial sphere has been covered so far. All known sources were found to lie close to the galactic

---

* An experiment in which the Sun can be likened to a "hot cathode", and the lunar surface to an "anti-cathode" of an "ion tube" of cosmic dimensions, whose glass walls (i.e. our atmosphere) enclose the observer rather than the apparatus.

The 200-inch reflector of the Palomar Mountain Observatory (1948).

plane (and $\pm$ 90° of the direction of the galactic center)
—a fact which strongly suggests that at least the majority
of these sources belong to our own Milky Way, and are
not extragalactic.* Moreover, their observed distribution
is so strongly reminiscent of that of the galactic Novae
as to suggest that most X-ray sources may be associated
with the remnants of supernovae. Indeed, Tau XR-1 was
located by observations of its occultation by the Moon
on July 7, 1964 as lying within one minute of arc of the
center of the Crab Nebula—which constitutes the remnant of the Chinese supernova of the year 1054 AD.
Another such source, Oph XR-1, fits (within the limits of
observational errors) the position of another supernova,
observed by Kepler in 1604; and a similar association
for several others is at least suggested. The brightest of
them all—Scorpius XR-1—appears to have been quite
recently (1966) identified with a faint 13-magnitude
stellar object, emitting a thousand times as much energy
in the X-ray domain as in the visible part of the spectrum.
On the other hand, the strongest known "radio stars" like
Cygnus A or Cassiopeia A were found to exhibit no de-
tectable signals in the domain of X-rays; but the reasons
why this is so can so far be only guessed at.

Are these the legendary "neutron stars" of pygmy size,
representing the last stage of a stellar collapse? X-ray

---

* With the exception of only one, which may be associated with
the Coma cluster of external galaxies. Extragalactic background of
soft X-rays (on wavelengths ranging between 44-70Å) was recently
detected by Bowyer, Field and Mack (1966) using Aerobee rockets.

observations of a lunar occultation of the Crab Nebula on July 7, 1964 (an event which will not recur for more than ten years) disclosed that the X-ray source in this object is very much smaller than the optical image of this nebula (see page 58) or the region emitting radio noise; and its emission amounts to about 10^{36} ergs/sec. In order to explain this as a black-body emission from the surface of a neutron star (of a size which such objects are expected to possess) it would, however, be necessary to assign to it a surface temperature of about 80 million degrees—which is far higher than a theoretical limit set up by the cooling effects of neutrino emission from the core. Such evidence as we now possess seems, therefore, to speak against the Crab source of X-rays being a neutron star. The absence of such an object inside the Crab Nebula does not, however, preclude the possibility that the Scorpius source constitutes such a star; for this latter source is not surrounded by any visible nebulosity, nor does it emit any radio noise. There may, indeed, be more than one kind of X-ray "star" around us in the Universe; and their studies from space will no doubt make important contributions to astrophysics in the future.

How far does the spectrum of such objects extend towards short waves, and which instrumental methods can be invoked for their studies? Before we attempt to answer these questions, let us consider briefly another type of vehicle by means of which such observations can be carried out in space: namely, the *orbiting observatories* which became available for use in the past five years, and which offer more effective platforms than rockets for sustained studies of celestial objects in space. Unlike the balloons, rockets in actual use in the past twenty years can lift scientific payloads into real empty space—not only the threshold of it—and open up the entire span of the electromagnetic spectrum of celestial bodies for closer inspection from their maximum altitudes. However—and this is essential—their time actually spent in free flight above the ionosphere is usually limited only to a few minutes; and while the instruments can sometimes be recovered after the fall by means of suitable parachute devices, the actual weight of such payloads is rather seriously limited to not more than a few hundred pounds. For more sustained observations it is necessary to extend the time spent in space to much longer intervals—even at the expense of the eventual loss of the entire payload, and the consequent need to telemeter all information to ground while the experiments are in progress. This has indeed been accomplished with the aid of orbiting observatories which we now wish to describe.

Observatories in Orbit

On October 4, 1957—a memorable date in the history of space exploration—our Earth acquired its first artificial satellite in the Russian Sputnik 1; and in the ten years which have elapsed since that time the number of artificial Earth satellites successfully launched into orbit has reached almost eight hundred. To give even a brief account of the magnificent effort behind this achievement is wholly outside the scope of this book; nor is it necessary to do so; for the large majority of these vehicles carried no telescopes in space. The number of missions which could be called astronomical was, in fact, limited so far to a relative handful of spacecraft which are listed in the accompanying Table I; and in what follows we wish to describe briefly their aims.

The primacy in this list belongs to the Russian Sputnik 3—successor to the pioneer Sputnik 1 (which decayed on April 1, 1958) and the dog-carrying Sputnik 2—which became the first scientific laboratory in orbit around the Earth. Of its total weight of 2,925 lb., its experiment instrumentation and power supply added up to 2,130 lb. Many experiments carried out aboard this spacecraft—magnetic field measurements, observations of the solar radiation in the UV-part of the spectrum, of cosmic rays as well as of the residual density and pressure in our atmosphere—have caused Sputnik 3 to be compared with the American multi-experiment Orbiting Geophysical Observatories (OGO) launched since 1964. It was with this spacecraft that the Russians have independently discovered the existence of the belts of charged particles surrounding the terrestrial globe (previously detected by van Allen with the aid of the American satellite Explorer 1).

Four spacecraft of the latter class were launched from the United States between 1964–1967 and carried instrumentation partly of interest for the subject of this book.

As shown schematically in the drawing on page 223, OGO's main body—68 inches long and 33 inches square—has two solar panels attached to it, with solar experiment packages mounted on each panel; two 22 ft. booms and four 6 ft. booms extend from the spacecraft to house magnetometers and other experiments that could be affected by disturbances generated in the body. Fully deployed, the satellite is 59 ft. long and 20 ft. across the solar panels. The total weight of the spacecraft of this class is close to 1,100 lb. Its communication and data-handling system can store up to 86 million bits of information, which can be transmitted at a rate up to 128,000 bits per second; the necessary electrical power is generated by 32,000 solar cells and nickel-cadmium batteries which furnish electrical energy up to the peak load of 560 watts. The altitude control system employs a horizon scanner with two Sun sensors to activate argon gas jets, and reaction wheels for Earth stabilization.

All four satellites of this class launched so far (see Table I) have been successfully injected in highly eccentric orbit inclined to the terrestrial equator by angles listed in column (7) of Table I, and attaining (for OGO's 1 and 3) apogee distances (column 6) up to one-third of the distance to the Moon. All four have remained operative up to the present time, and transmitted a wealth of scientific data.

Of twenty experiments aboard OGO-1, those of primary interest to us in this book have been the measurements of the solar radio noise and bursts (conducted by the University of Michigan), of solar cosmic-ray protons and X-rays (University of California) as well as the mapping of the cosmic radio noise over the sky (University of Michigan). Not everything went well, to be sure, with this spacecraft; for two of its booms failed to deploy, and one obscured the horizon scanner. As a result, the altitude of the spacecraft could not be orientated with respect to the Earth, and OGO-1 remained spin-stabilized at five revolutions per minute. Nevertheless, useful data (so far largely unpublished) are still being received from sixteen of the twenty experiments aboard; and may continue to be received for many years to come.

OGO-2, launched on October 14, 1965 in a near-polar orbit around the Earth, carried a similarly directed instrumental payload, aimed at studies of the terrestrial space environment. Unlike the preceding spacecraft of this series, OGO-2 deployed its booms satisfactorily; but its horizon scanners drifted on occasion off the Earth lock as the spacecraft passed over the equatorial regions (apparently due to the effects of a cold air mass above the equator). The need of repeated Earth acquisition depleted the gas jet supply ten days after the launch; but

A telephoto picture of the Earth, taken by Orbiter 5 from lunar proximity on 8 August 1967. The outlines of the principal geographic features of a large part of our planet are clearly seen on the photograph. (By courtesy of NASA and of the Boeing Company.)

the spacecraft was able to orientate its solar panels towards the Sun, and nutate slowly around its long axis. Nineteen of its twenty experiments are returning useful data which are being currently analyzed.

OGO-3 was instrumented (among other tasks) to measure the rigidity and charge spectrum of the cosmic rays of solar as well as galactic origin, in order to study the modulation mechanisms of cosmic radiation in the solar system (University of Minnesota); to continue the mapping of the radio noise over the sky (including solar noise; University of Michigan) or to determine variations in solar X-ray flux at wavelengths between 0.5 and 60Å (U. S. Naval Research Laboratory). The entire spacecraft is operating as planned, and continues to return data from all twenty-one experiments aboard. Simultaneous measurements by OGO-1 and OGO-3 at different locations within the magnetosphere, together with OGO-2 measurements in low Earth orbit, provide opportunities for studies of the terrestrial space environment undreamt of before the advent of these spacecraft.

OGO-4, launched on July 28, 1967 into a near-polar orbit, represented another space vehicle for a study of solar-terrestrial relations during the time of increasing solar activity. Twenty experiments aboard (18 of which are providing useful data) are directed mainly towards studies of particle activity, aurora and airglow, geomagnetic field, and electromagnetic ionization sources; but none of them belong properly to the domain of optical astronomy. OGO-5, the latest member of this family, was successfully launched in its polar orbit on March 4, 1968, with similar instrumentation; and OGO-6 —the last remaining member of this family of spacecraft yet to be launched—will follow a similar orbit.

The next family—the first to be launched—of Orbiting Solar Observatories (OSO), is of greater interest to the students of astronomy. The OSO's (see page 223) have been designed primarily as stabilized platforms for solar-orientated instruments; the major program objectives being the measurements of solar UV, X-ray and γ-ray radiation from the Sun, and its variation from the Sun, radiation and its variation with the time.

The OSO spacecraft consists of two major components: a spin-stabilized "wheel" section 44 inches in diameter, and a fan-shaped "sail" section containing the experimental equipment together with an array of 1,860 solar cells. The overall height of the spacecraft is 37 in.; and its weight, 440–90 lb. with a minimum of 175 lb. for the experimental payload. The wheel section consists of nine wedge-shaped compartments—four for spacecraft electronics and five for experiments not required to point continuously at the Sun. Three fiberglass balls extending on arms from the wheel section contain nitrogen for the spin control system by means of gas jets. The wheel section, whose plane of rotation includes the Sun, is designed to spin at thirty revolutions per minute. The sail section (above the wheel) can lock on the Sun with an accuracy of one minute of arc during the daylight part of each orbit by means of an altitude control system employing electric motors and gas jets.

The first such Orbiting Solar Observatory was launched by NASA with success on March 7, 1962; and during its seventy-seven days of almost faultless operation it transmitted to the Earth from its lofty vantage-point between 340 and 370 miles above the Earth nearly 1,000 hours of data. It observed more than 140 solar flares—those "outbursts of temper" of our mother Sun, which are accompanied by intensive (albeit local) emission of ultraviolet radiation and X-rays—it mapped the sky in γ-radiation, and examined energy particles in the low-altitude van Allen belts. Significantly, it also found discrepancies between different manifestations of solar activity observable above the atmosphere, as compared with ground-based observations.

The OSO-2, launched on February 3, 1965, carried eight experiments aboard, which included a group of X-ray solar monitors and pinhole cameras operative between 2 and 60Å (Naval Research Laboratory); a white-light coronograph (NRL); and a spectro-heliograph operating at wavelengths of $\lambda = 304$Å (i.e. Lyman-α line of ionized helium), 584Å (the corresponding line of neutral helium), and 1,216Å (neutral hydrogen), designed likewise by NRL. Among experiments mounted in the wheel section of OSO-2 there was a photometer to measure the intensity and degree of polarization of the zodiacal light around the Sun (University of Minnesota), and an ultraviolet stellar and nebular spectrophotometer operating in the wavelength region of between 1,500 and

3,300Å (NASA Goddard Space Flight Center).

The scientific data obtained aboard OSO-1 and OSO-2 were telemetered to Earth by two independent tape-recorders and transmitters to minimize the noise. During 90 min. of each OSO's 96 min. orbit information reported by different instruments was being fed into a tape-recorder. In the course of five out of the remaining six minutes, one of the five NASA ground stations (Fort Meyers, Florida; Blossom Point, Maryland; Quito, Ecuador; Lima, Peru; Santiago, Chile; and Woomera, Australia) monitoring the OSO's commands the spacecraft to transmit the data. This transmission can be accomplished at a rate of eighteen times faster than that of the recording. As, moreover, OSO transmits its information, it simultaneously wipes its tape clear so that when the transmission ceases, in the remaining one minute of the orbit the spacecraft can be reset to record new data.

The next spacecraft of this series—OSO-C—was launched by NASA with an instrumentation similar to that of its predecessors on August 25, 1965; but failed its injection into circumterrestrial orbit because of a premature ignition of the third stage of the carrying rocket. No other spacecraft of this class were launched in 1966; but 1967 witnessed the launch of two more (OSO-3 and OSO-4), instrumented mainly for X-ray photography of the solar disc in spectral domains between 3 and 8Å, 8 and 20Å and 20 and 50Å; in addition to high-resolution spectural studies between 300 and

1,300Å.

OSO-3, launched on March 8, 1967, was a repeat of OSO-C which failed to orbit on August 1965. This spacecraft had nine experiments aboard, designed to study the structure, dynamics and chemical composition of solar outer layers in the X-ray, UV, as well as the visible domain of the spectrum.

OSO-4, launched on October 18 of the same year, provided us with the first (digitally transmitted) pictures of the Sun in the light of the ionized oxygen line Ovi at $\lambda = 1032$Å, of the Lyman continuum at $\lambda = 987$Å, and of the ionized magnesium Mgx at $\lambda = 625$Å. These "digital spectroheliograms", taken between October 25–27 and reconstituted at the Harvard Observatory, are reproduced on page 157 together with a conventional Hα spectroheliogram taken at the same time from the ground at the Sacramento Peak Observatory in New Mexico.

Further launches can be expected in the future; for (owing to their relatively moderate orbital altitudes) the mean lifetime of an individual OSO is not likely to exceed one to two years before its eventual decay; and a complete monitoring of at least one entire solar cycle of eleven years will require successive co-operation of several additional OSO's.

The solar radiation and its doings are, after all, very important to us in a very real sense. For not only does the Sun provide us (and has done so from time immemorial) with the warmth of daylight to mitigate the

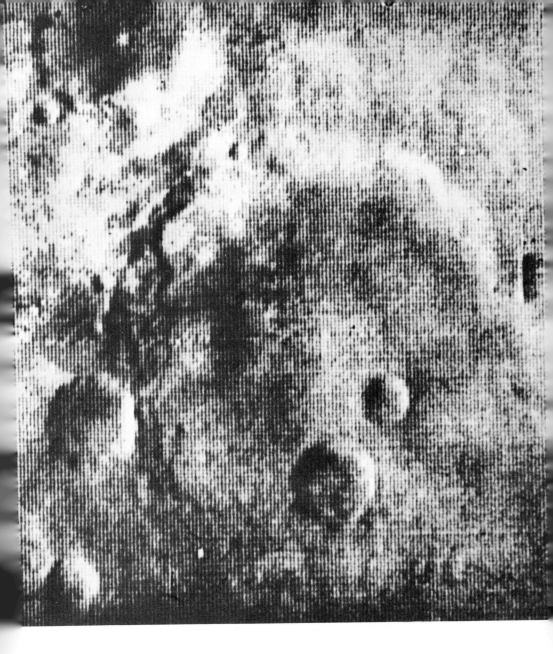

A photograph of the Martian surface, obtained by the U.S. spacecraft Mariner 4 on 14 July 1965 from a distance of 12,400 km. The field of view is 270 × 240 km in size, and the largest crater (with incomplete walls) is about 170 km across.

rigors of starry nights, but it controls all our weather—rain, snow, winds, even ocean currents are but different manifestations of a "heat engine" driven by the solar energy output. But more than that: virtually all our other sources of energy—water power, coal, oil or other combustible fuel such as wood—derive in various ways from the sunshine: water power from the ability of solar heat to evaporate water into clouds, from which it comes down as rain to make our rivers flow back into the seas; while wood, coal or oil represent nothing but contributions of sunshine to the growth of plants or animals—recent or fossil. In point of fact, all kinds of energy which we use in daily life goes back, in one way or another, to the Sun—with the sole exception of the nuclear energy released in recent years by human hand by induced fission of heavy elements; for these antedate the Sun in age and owe their origin to other stars which ended up their lives before the Sun was born.

As to the origin of sunlight proper, we know now that, physically speaking, the Sun constitutes nothing but a slowly smoldering thermonuclear reactor, burning in its deep interior hydrogen into helium, and releasing the balance of the energy in the form of radiation. Only a small part of the Sun's mass participates in this process; the rest serves as a shield of the central reactor, and sees to it that the light born as X-rays is gradually transformed into such light as will eventually leave the surface to travel freely in space; and a tiny bit of it may eventually be intercepted by a solid speck known as the Earth, to tell its inhabitants that the Sun is shining.

It is indeed a tiny bit of total radiation from the Sun which falls on the Earth and is absorbed to a variety of purposes; for the Sun as a whole emits (in all directions) more energy per second—about 4×10^{33} ergs, corresponding to 4×10^{23} kilowatts—than all people living on this Earth have used, or consumed, since the dawn of our civilization. Most of this energy comes to us in the form of what we call "visible light." Rockets and satellites operating above the atmosphere discovered, however, in recent years that during localized "outbursts of temper" the Sun emits also bursts of X-rays (see page 157) which would be lethal to life on the Earth without the additional protection of the atmosphere; and would be equally lethal in space or on the Moon without adequate protection. One of the sidelights emerging from the current space work is the realization that "sunbathing" in space could be a very risky business in more than one sense!

Unlike the Sun, all other stars are so far away from us that their radiation constitutes no risk. However, their study from space should be no less rewarding for the range of the phenomena which they exhibit. In order to initiate it in a systematic manner, a new class of Orbiting Astronomical Observatories (OAO) has recently been added to the OGO's and OSO's; and the first one of them has already been launched into orbit.

The primary aim of the spacecraft of this class is to undertake systematic observations of celestial bodies over

OVER.

After sunset on the Moon—the streamers of the solar corona as recorded by Surveyor 1 above the lunar horizon in June 1966. (By courtesy of JPL, CIT.)

the entire sky in those parts of their spectra which are inaccessible from the ground because of atmospheric absorption, and which require observing platforms stabilized in space for protracted intervals of time. Experiments already selected for the program include a survey of the brightness of some 50,000 hot stars (down to approximately eighth apparent magnitude) in three spectral domains of the ultraviolet; photometric measurements of ultraviolet spectra of a smaller number of selected stars and emission nebulae; and investigation of the absorption characteristics of interstellar gas.

A typical spacecraft of this class—(for a drawing see page 223)—consists of an octagonal frame 118 inches long, and the distance between opposite flats of the octagon is 80 in. The spacecraft is, therefore, almost 10 ft. high and 7 ft. wide. A pair of solar paddles (112 sq. ft. in size) are mounted on opposite flats of the cylinder at an angle of 56 degrees, which can be folded prior to injection into orbit, and opened up later to give this spacecraft a wing spread of 21 ft. Concentric with the octagonal frame is a circular cylinder, in which the experimental equipment is mounted. The diameter of this cylinder is 48 in., and it spreads through the entire length of the spacecraft. The gross weight of a fully instrumented OAO should be close to 4,000 lb.—making it a much heavier spacecraft than either OGO or OSO— of which about 1,000 lb. will be consumed by the actual experiments, and the balance by the basic structure and subsidiary subsystems.

The latter will include both coarse and fine reaction wheels, nitrogen gas jets and magnetic torque coils to stabilize the spacecraft for guiding to an ultimate accuracy of 0.1″. As to the communications, the OAO's should be in contact with their monitoring stations on the ground (located at Santiago, Chile; Quito, Ecuador; and Rosman, North Carolina) for approximately 10 per cent of their operating time (i.e. about 10 min. out of each 100-min. orbit), and will carry memory systems capable of storing and playing back over 200,000 bits of information accumulated per orbit. Three independent transmission limbs will operate at frequencies close to 136 MHz. (two), and 400 MHz., and will be powered mainly by more than 74,000 solar battery cells mounted on the solar paddles. In daylight these batteries can supply up to 400 watts of electrical power for different needs of the space station. During the night (i.e. when the OAO is in the Earth's shadow) the power is supplied by a nickel-cadmium storage battery, kept charged by the solar batteries during periods of sunshine.

The first OAO, launched on April 8, 1966, carried two major experiments aboard: namely, the University of Wisconsin's ultraviolet broadband spectrometer (nebular photometer), with two additional scanning spectrometers to study selected stars and nebulosities; and the X-ray and γ-ray telescopes for the NASA's Goddard Space Flight Center, the Massachusetts Institute of Technology and the Lockheed Aircraft Corporation.

The primary objective of the Wisconsin experiment

(conducted by A. D. Code and T. Houck) was the determination of stellar energy distribution in the spectral region between 800 and 2,900Å, and the measurements of emission line intensities of diffuse nebulae. Secondary objectives included investigations of the intrinsic variability of stellar energy distribution, of the law of interstellar reddening, etc. There are essentially three basic systems for measuring the intensity of radiation from celestial sources, in selected wavelength regions, envisaged by the Wisconsin experiment. That for measurement of stellar energy distribution consists of four identical telescopes, consisting of 8-inch off-axis parabolic mirrors of 32-inch focal length, with a photo-electric photometer mounted at the prime focus and equipped with a five-position wheel carrying filters centered at different passbands about 300Å in width. The nebular photometer uses an on-axis parabolic mirror of 16-inch diameter and 32-inch focal length, with the photometer mounted likewise at the prime focus; and the filter wheel transmitting around the wavelength of 200, 2,500, 2,800 and 3,300Å. The detector is an EMI photo-multiplier with caesium-antimony cathode; and the electrical signal generated by it is stored and eventually transmitted to the ground.

The Goddard experiment—the second to share the ride on the first OAO in April 1966—was primarily intended to obtain absolute spectrophotometric data on selected stars, nebulae and galaxies. For normal stars, the program should include a study of the energy distribution in the continuum, and intensity measurements of strong emission lines. For particular stars (such as variables of the type of β Canis Maioris, T Tauri, or Wolf-Rayet stars) the variation of light with the time should be measured; and the laws of interstellar absorption as well as emission investigated. The spectra of the emission and reflection nebulae, the spectral energy distribution of near-by galaxies, and the magnitude of the hydrogen Lyman-α red shifts are also among the observational objectives.

The equipment with which the Goddard scientists plan to undertake this program is basically a 36-inch reflector of the modified Ritchey-Chrétien design (the primary mirror of which is a concave quasi-hyperboloid, and the secondary a convex-hyperboloid). Owing to the limits of room aboard the spacecraft, the primary mirror is extremely fast (f/1.6), and with a primary-secondary spacing of 42-inch the effective focal length of the system amounts to 190-inch. This instrument should be capable of observing stars as faint as the tenth magnitude, and of scanning the spectural range between 1,050 and 4,000Å with a spectural resolution of 2, 8 or 64Å. High-energy cosmic γ-rays were the subject of an experiment from M.I.T.; while Lockheed experiments set out to study soft stellar X-rays with the aid of gas proportional counters.

On Friday, April 8, at 2:56 p.m., Eastern Standard Time, an Atlas-Agena rocket roared upwards from Cape Kennedy to place the OAO-1 in a near circular orbit 500 miles above the surface of the Earth; and for the first time in history a 36-inch telescope (as large as that previously lifted to balloon altitudes by Stratoscope II) was successfully injected into circumterrestrial orbit well above the atmosphere. For a time it appeared that astronomers would at last be able to operate a good-sized telescope in real space, and obtain unique observations of the stars in the UV and X-ray domains of the spectrum. But, unfortunately, the next day battery trouble had begun to develop, as well as motor trouble with the spacecraft's command clock; and on Sunday NASA's statement disclosed that "Overheating of the OAO primary battery has resulted in complications leading to degradation of the power supply from all three batteries aboard the spacecraft, and telemetry signals are no longer being received." Efforts to overcome the problem were unsuccessful, and the first OAO mission has been lost to orbit around the Earth with its precious cargo—mute and silent—for many years to come.

Several other OAO's are, however, currently in preparation and the next one is scheduled for launch in the latter part of 1968 with—we hope—better success. Thus the Smithsonian Astrophysical Observatory (F. L. Whipple and R. Davis) in an effort called "Project Celescope", is preparing four large-aperture television cameras fed by Schwarzschild-type reflectors of 12-inch aperture of 24.9-inch focal length, permitting adequate imagery over a field of 2.8° (1.2-inch) in diameter. The aim of these instruments should be to map the entire sky in four ultraviolet spectral bands around 1,400, 1,500,

2,300 and 2,600Å; and the sensitivity of the system should be sufficient to allow observations of some 50,000 stars down to the eighth visual magnitude. The mapping should also include nebulosities whose surface brightness exceeds eighth magnitude per square minute of arc (which should include, for example, the Orion Nebula and some brighter planetary nebulae). In addition the Princeton University Observatory (L. Spitzer and J. B. Rogerson) are preparing to orbit a 32-inch reflector of 98-inch focal length, to study early-type stars in the range between 700 and 3,200Å with high spectral resolution between 0.05 and 0.4Å.

Prospects for the Future

In the preceding parts of this chapter we introduced to the reader the two principal vehicles of space astronomy at the present time: namely, the rockets and Earth satellites. How do they compare in performance of their respective tasks? The figures we quoted already leave no shadow of doubt that, since the commencement of this decade, the satellites have outpaced the rockets to such an extent that any direct comparison becomes almost invidious. It is true that rockets can lift in their warheads limited scientific payloads well above the ionosphere—and were historically the first vehicles to do so —but only for minutes at a time; and thus cannot secure more than glimpses of the celestial scenery which surrounds us in space. Satellites in circumterrestrial orbits can be made to attain arbitrary altitudes (much higher

The capsule of the Russian spacecraft Luna 9, which landed softly on the lunar surface on 3 February 1966. (By courtesy of the USSR Academy of Science.)

than are possible by rockets) and—what is much more important—operate at them for incomparably longer intervals of time.

A satellite whose instrumental package remains operative for one year (and this has nowadays become a common achievement) will accumulate some 4,300 observing hours if it revolves in close proximity to the Earth, and almost double that amount in wide orbit (in which it does not undergo occultations in the terrestrial shadow at each revolution). It has been estimated by NASA that about 12,000 sounding rockets, operating perfectly, would have been required to acquire the same amount of information as did OSO-1 during its seventy-seven day active career in space; and the cost of the launching of so many rockets would have been one hundred times as large. In point of fact—all considered—observing time aboard a satellite costs considerably less than it does aboard either a rocket or stabilized balloon platform. To launch a single balloon or rocket is, to be sure, much less expensive than the launch of a satellite; but observing time then comes in trickles, while satellites furnish it at a wholesale price.

It is true that there are other advantages to rockets which satellites would be hard pressed to match. Like balloons, rocket experiments possess in general much shorter lead-times and (within the limits of their allowable payloads) may offer more versatility to the experimenter. Also, at least a part of the results secured by them can sometimes be directly recovered; and their acquisition does not, therefore, have to rely wholly on telemetry. The relative short lead-time of rocket launches makes them particularly suitable for observations of transient phenomena—such as solar flares, or occultations of different objects by the Moon—though a satellite in orbit instrumented for the same kind of observations can do so equally well or better.

All these points in favor of the rockets cannot, however, outweigh the tremendous advantages of orbiting satellites in providing unlimited amounts of observing time from almost arbitrary vantage-points in space. Although their operations require relatively large initial outlay, there is no doubt that they can supply unrivalled observing opportunities at less cost per unit time than any other means at our disposal at present—and will continue to do so until observing bases can be set up on the Moon. Rockets may retain more of permanent usefulness in geophysics (where they can be deployed to advantage for studies of transient changes of the environment); but in astronomy their role will, in the years to come, be largely limited to act as boosters for orbiting spacecraft.

With these prospects in mind, which types of instrumental developments in payloads can one envisage for the future? There is no doubt that the greatest single contribution of space astronomy so far has been the discovery of X-ray emissions from the Sun, and X-ray sources among the stars; and the quest is currently being pressed towards shorter wavelengths in the domain of the γ-rays.

It should be stressed that as yet no ground-based, balloon-borne or satellite-borne experiment has provided compelling evidence for more than upper limits of cosmic γ-rays at wavelengths shorter than 0.01Å (i.e. corresponding to photons with energies greater than 1 MeV). Moreover, such limits as have been established extend uniformly over the entire sky, and indicate the presence of no discrete cosmic γ-ray sources.* Even these upper limits are, however, useful in imposing corresponding limits on some of the large-scale properties of energetic particles in interstellar (and intergalactic) space.

Thus the upper limit for the intensity of diffuse cosmic γ-rays, established by Explorer 2, is about twenty times as large as one would predict for γ-rays produced by cosmic-ray collisions with interstellar hydrogen atoms (through the decay of neutral pions); and other processes of γ-ray production (such as magnetic or collisional "brems-strahlung", or Compton scattering of relativistic electrons and positrons; or de-excitation of heavy atomic nuclei following neutron capture) which could produce line emissions in the γ-ray spectrum; but no indications of them have been established so far from observations.

Owing to their high energies, γ-rays can propagate through interstellar or intergalactic space with almost no

* Unless one be identical with the well-known X-ray source Cygnus XR-1, in the neighborhood of which the University of Rochester physicists reported in 1966 a marked increase in γ-ray intensity registered by their balloon flights to an altitude of 120,000 ft.

The American spacecraft Surveyor 1, which soft-landed on the lunar surface on 2 June 1966. (By courtesy of the Jet Propulsion Laboratory, California Institute of Technology.)

hindrance by the matter which they encounter on the way, and thus permit observations of objects which may be obscured to other forms of electromagnetic radiation. In the future, γ-rays may indeed provide us with new means for probing the structure of the center of our Galaxy, or of communicating with the most distant parts of our Universe. On the other hand, their penetrability through our atmosphere also increases with diminishing wavelength; and while to register cosmic X-ray photons of one Ångstrom wavelength (carrying energy of about 12.5 keV) it is still necessary to ascend to an altitude of not much less than 60 miles above the ground, γ-ray photons with energies greater than 100 MeV can penetrate all the way down to the ground—thus opening for us a third "window" (in addition to that at optical, and radio-frequencies) in which our atmosphere becomes once more transparent to radiation from outer space. But a view through this last window has not shown us anything so far; for observations made with appropriate ground-based counters failed to register the flux of any photons of cosmic origin in this energy range.

What kind of instrumentation may be needed in the future for further advances in X-ray or γ-ray astronomy? In the domain of X-rays, directional photon counters are already giving way to focusing X-ray optics (consisting of a reflecting surface on which X-rays impinge at small angles and are totally reflected). Such optics can be made essentially achromatic, and it does not alter the state of polarization of the impinging radiation; though two or more reflections may be necessary to remove first-order aberrations. Telescopes of this type of quite moderate dimensions (80-inch focal length) could permit in one hour of observation to image X-ray sources 0.01–0.001 times less intense than the Crab Nebula, with a resolution of 5 sec. of arc; and telescopes with a focal length of 30 yds. or more should increase the sensitivity to the point where sources with an intensity about one-millionth of that of the Crab Nebula may be detected. Side by side with studies aiming at mere detection or mapping of the positions of X-ray sources in the sky, grazing-reflection spectographs will be used to search for indications of line spectra of such sources: and other instruments will be employed to investigate possible polarization of their light.

All these systems have been (or can be) developed for light whose wavelength is not much less than one Ångstrom. At substantially higher frequencies, the molecular structure of matter will give out on us; and the γ-ray astronomy will call for instrumental equipment even more different from the conventional outfit of the astronomers. Already in the domain of photons with energies between 0.1 and 30 MeV no focusing is possible any more; and one must use detectors which depend on the absorption or scattering phenomena. Omnidirectional sodium-caesium scintillation counters were already used aboard OSO-1; and in the future such spectrometers may be used in pairs to gain angular resolution.

Above 100 MeV the greatest single difficulty for experimenters to overcome is the noise—both cosmic and instrumental. In this energy range the total flux of primary radiation is probably not more than a few quanta per square meter per second reaching us from a solid angle of one steradian of the sky. Surprises may be in store for us in the future; but this estimate represents the threshold of sensitivity necessary for the detector. The primary flux of charged particles of cosmic origin in the same energy range is at least a thousand times as large; and their interaction with the atmosphere (or, in space, with the walls of the vehicle, or the detector itself) produces secondary γ-rays which (except for possible directionality of the radiation from cosmic point-sources) may be hard to distinguish from the primary γ's. More than in any other branch of observational astronomy, the investigator of stellar γ-rays will be faced with the problem of the detector noise; but no doubt he will cross this bridge when he comes to it in some manner; impressive and continuing advances in this field of instrument technology give valid grounds for optimism in the future.

Before concluding this chapter we wish to make one retrospective comment concerning the achievements of satellite astronomy which we had an occasion to report in this chapter. The reader noticed undoubtedly that the large majority of advances which we described have come from American sources. Does it mean that the parallel contributions of the Russians to the current space effort have been ignored, or proved insignificant? This is indeed far from being the case.

As is well known, the Russians were the first ones to launch into orbit, not one, but two spacecraft (Sputnik 1 and 2) in the autumn of 1957 before the first American satellite (Explorer 1) was successfully launched on January 31, 1958. Since that time, up to the end of 1967, a total of 486 spacecraft were launched into geocentric orbits from the United States (of which 252 are still in orbit)—in contrast with 234 launchings by the U.S.S.R. during the same period (of which 56 continue to orbit). By the number of launchings, the U. S. has by now overtaken the U.S.S.R.; however, by the weight of the payloads carried by them into space, the U.S.S.R. is still clearly in the lead. There is, in particular, nothing on the U. S. side as yet to compare with the Russian mighty 13-ton Proton spacecraft orbiting high-energy physics laboratories around the Earth.

But—and this is essential—the bulk of the Russian space effort has so far been directed even more than in America to studies of the physical environment of the Earth in space (i.e. towards particle physics) and to investigation of life-support systems rather than toward astronomical observations of other celestial bodies; and as such it is largely outside the scope of our book. This overall orientation reflects the preponderance of the physicist and biologists over astronomers in councils planning space activities in both the U. S. and the U.S.S.R.

The astronomers have, to be sure, no grounds to

complain of professional discrimination; for although they can clearly benefit more than any other group of scientists from these new avenues of research, their traditionally more conservative "esprit de corps" has held them so far back at least as much as their smaller numbers. Three hundred years ago it took our astronomical ancestors more than half a century to come to terms with that newfangled invention of the telescope; and although there are indications that it may now take less than that time before we collectively embrace spacecraft as a legitimate instrument of astronomical research, particle physicists and biologists have already forestalled us in some of the early rounds of the game. Far from begrudging them this position of temporary privilege, we should rejoice in the achievements brought about by their ingenuity; for science is really one—and of the book of space astronomy we have seen so far only the first few pages.

5 · Into Deep Space: Lunar and Planetary Probes

While the rockets were popping up in the ionosphere like fireflies, and satellites began circling the Earth in increasing numbers, a new class of spacecraft commenced to be launched by human hand to explore greater depths of space and pay close introductory calls on our nearest celestial neighbors—the Moon, Venus and Mars.

Efforts to launch spacecraft which could disengage themselves from the gravitational field of the Earth go back to 1958, when the American rocket Pioneer 1, launched on October 11, reached a distance of 70,717 miles from the Earth; while Pioneer 3 (launched on December 7, 1958) attained only a slightly smaller distance, 63,580 miles. It was the Russians who provided the Sun with its first artificial planet Luna 1, which was launched on January 2, 1959, and by-passed the Moon at a distance of a mere 3,728 miles. The Americans followed up with a second artificial planet, Pioneer 4, about two months later (March 3). The Russian Luna 2, launched on September 12, made history by scoring the first direct hit on the Moon's face on September 14 after a flight of 63½ hours; and not quite a month later Luna 3 (launched on October 4) performed the first circumnavigation of the lunar globe and unveiled to us for the first time the principal features of the topography of its far side.

Luna 1 and Pioneer 4 were joined in their heliocentric orbits by Pioneer 5 on March 11, 1960; and this latter feat closed the glorious ledger of space exploration of the sixth decade of this century; but the subsequent years were no less rich in accomplishment. On February 12, 1961 the Russian Venus 1 probe was launched; and next year not less than four space probes (Ranger 3 on January 26, 1962; Mariner 2 on August 26; Ranger 5 on October 18 and the Russian Mars 1 on November 1) have joined the growing artificial interplanetary family of man-made asteroids revolving around the Sun. In 1963 no new members were added; but the year 1964 witnessed the launching of four new such bodies—two Russian (Zond 1 on April 2 and Zond 2 on November 30) and two of American origin (Mariners 3 and 4 launched on November 5 and 28)—and in 1965 no less than five additional deep-space probes were launched: four from Russia (Luna 6 on June 8, Zond 3 on July 18, Venus 2 and 3 on November 12 and 16) and one from America (Pioneer 6 on December 16), to bring the sum total of all

Stratoscope II in pre-flight checking facility. (By courtesy c the Perkin-Elmer Corporation.)

artificial spacecraft launched by human hand into helio-centric orbits to seventeen. As only one new launching of similar spacecraft was made in 1966 (Pioneer 7 on August 17) and two in 1967 (Venus 4 and Mariner 5 on June 12 and 14, respectively), the total now stands at twenty; though probably not for very long. That so many interplanetary spacecraft were launched in 1964 and 1965 or 1967, and so few in 1963 or 1966, is largely due to the relative configurations (i.e. oppositions or conjunctions) of the principal objectives of this celestial target practice: namely, the planets Venus and Mars.

To give even a brief account of the individual missions of all these artificial asteroids would take us too far afield in a book devoted to the telescopes in space; for our main topic directs our attention to the telescopic exploration of three principal targets beckoning to us across the intervening gaps of interplanetary space: namely, the planets Venus and Mars, and—above all—our nearest celestial neighbor: the Moon.

The Moon—a celestial globe one-quarter the size of the Earth, and revolving around us in a mildly eccentric ellipse at a mean distance of 240,000 miles (60.27 times the Earth's equatorial radius)—has from time imme-morial been the target of human aspirations to set foot on the surface of this nearest celestial body. From the entrancing visions of Kepler's *Dream,* through William Herschel to Jules Verne and his followers, this motive has haunted human imagination in various forms.

Many a reader of this book may remember from his

youth Verne's famous fictional story of an attempt by the American cannoneers to reach the Moon by means of a manned projectile fired from a gun—the legendary Columbiad of the Baltimore Gun Club. Today, a century later, it is easy to see all that would have gone wrong with it in practice. Perhaps the gravest mistake Verne committed was to place men in his projectile; for should President Barbicane and Captain Nichols have really been induced by Michel Ardan to enter with him the hollow shell designed by the worthy T. J. Maston, the fate of all three intrepid travellers would have been gruesome beyond description. Suffice it to mention that, in order to accelerate a projectile to the velocity of 6.8 miles/sec. (necessary for the escape from the gravitational field of the Earth) inside a gun barrel of the dimensions of the Columbiad, it would have been necessary to expose the astronauts to an initial acceleration of almost 30,000 g's—which no living organism could possibly survive for a fraction of a second.

And not only the travellers inside, but also the projectile itself would have been subjected to a crushing strain between the expanding gases of the explosive charge, and the resistance offered inside the barrel of a body pushing it out with a velocity of more than thirty times that of sound. The hollow projectile would have been squashed in between to a cake, and heated to more than a thousand degrees to become a crematorium for its crew within a few seconds of their flight, which would have ended not far from its place of origin—to the

Photographs of the Sun on May 20, 1966, taken in the light of the hydrogen line Hα (λ6563Å) in the red part of the spectrum (top) and in the X-ray domain between 9-20 A (bottom.) This latter photograph was obtained with the aid of a rocket-borne glancing-incidence telescope of focal length 62 cm., operating at a focal length of f/44, and attaining a resolution of about 7 seconds of arc. Note a strong localization of X-ray emissions in the enhanced regions of bright faculae. (After Underwood and Muney, 1967.)

intense disappointment of countless spectators and to the horror of their friends. Although, from a technical point of view, the event envisaged by Verne would thus have ended in a grim fiasco, in one respect the great French novelist did anticipate closely the future: namely, by placing the scene of his narrative—Tampa Town in Florida—only a little more than 100 miles from Cape Kennedy, where a major part of the modern sequel to Verne's story is being enacted this decade—a sequel in which giant guns of the past have given place to their modern successors—the rockets.

Why go to the Moon? Why has it become so far a target for more space missions than any other celestial body? Not only because "it is there", or because of proximity alone; but also because such studies are of surpassing scientific interest in a much wider context. The Moon is a very old body, and has probably been a close companion to our Earth for several milliards of years. A permanent and virtually complete lack of any air or liquid on its surface makes it, moreover, certain that the most part of its composite fossil record must be very ancient—its oldest visible landmarks being probably not far removed from the time of the origin of the solar system. On the Earth (or any other planet surrounded by an atmosphere and hydrosphere) all landmarks of comparable age would have been obliterated by the disturbing action of air and water eons ago. However, as any changes on the Moon caused by other kinds of erosion (seismic, sputtering light) can proceed only at an exceedingly slow rate, its present wrinkled face may still bear traces of many events which have taken place in the inner precincts of the solar system at times going back to its origin; and if so, their correct interpretation holds indeed a rich scientific prize.

A more complete information concerning lunar space missions undertaken from 1959 through the first half of 1968 is listed in the following Table II, in which the entire set of data has been divided in the following four groups: (a) lunar fly-by's; (b) hard landings; (c) soft landings; and (d) orbiters.

Of group (a), Luna 3 (see page 235) was an artificial Earth satellite of a period close to 16.2 days and reaching an apogee distance of 291,440 miles, ceased to exist on April 20, 1960, when it met its fiery end by air resistance in our atmosphere; but all other members of this group have gone into heliocentric orbits of unlimited lifetime. This includes the Russian Zond 3 (see page 197) of July 1965, which took improved photographs of the far side of the Moon—the best ones we had before those taken by the American Orbiter 1 in August 1966.

The selenographic positions of the impact points of the hard-landers of group (b) are shown on the accompanying page 89 (with the exception of that of Ranger 4 which landed on the Moon's far side); and so are those of the soft-landers listed in group (c). The Lunas 4 and 6, or the last five entries in group (b) represented, in fact, attempts at soft landings which failed because of different types of instrumental malfunctioning. Only three spacecraft listed in group (*d*) still continue to orbit the Moon at the present time; but their lifetimes are limited by the cumulative effects of lunar perturbations. In particular, Orbiter 1, launched on August 10, 1966, met its end on October 29 (when it was instructed from Earth to commit suicide in order not to interfere with the signals of its successor); Orbiter 2 ceased to function on December 6; and the last one crash-landed on the lunar surface on January 29, 1968.

Fly-by Spacecraft

The principal contribution to lunar studies by fly-by spacecraft has been to provide, in 1959 and again in 1965, the first photographs of the far side of the Moon, which has been inaccessible to direct observation from the Earth since the Moon's axial rotation has become synchronized with its revolution as a result of the tidal friction continuously operating in the Earth-Moon system; and this synchronization must have been virtually complete throughout the most part of our geological past.

It is true that the optical librations of our satellite (due to the eccentricity and inclination of its relative orbit) enable us to see appreciably more than one-half (approximately 59 per cent) of the entire lunar surface from the Earth at one time or another; only 41 per cent being permanently invisible. This was so until October 1959, when the cameras aboard the Russian spacecraft Luna 3 unveiled for us the main features of a major part (28 per cent) of the far side of the Moon (see pages 150 &

161). The relative position of Luna 3 at the time of photography was such that about 13 per cent of the lunar far side was invisible from the spacecraft and remained uncharter until July 20, 1965, when another Russian space probe (Zond 3) succeeded (see page 161) in recording all but a small fraction of the remainder.

The Russian photographic experiment aboard Luna 3 in 1959 performed in the small hours of October 7 is still in too good a memory to need much repetition in this place. The optics of this spacecraft consisted of two lenses of 36 and 52 mm. aperture, and 20 cm. and 50 cm. focal length, by which a number of frames were taken with different exposures (of the order of 10 millisec.) on a 35 mm. film. After the exposures were taken, the film was automatically processed aboard and scanned for telemetry to Earth where the original image was reconstituted by television.

At the time of the photography Luna 3 was between 40 and 42 thousand miles behind the far side of the Moon; and from its vantage-point the Moon would have appeared as a disc of an angular diameter of close to 3 degrees (i.e. about the same as the Moon would appear to us from the Earth through an ordinary field-glass). The ground resolution of these photographs—one of which is reproduced in a photograph on page 150 was about 15–20 miles on the lunar surface; and their principal scientific contribution has been the discovery that the Moon's far side consists predominantly of mountainous ground and contains very few maria—a discovery amply confirmed by subsequent space work.

Perhaps nothing illustrates the rapidity of acquisition of new knowledge in the field of lunar studies more eloquently than the progress in our acquaintance with the topography of its far side. While the Russian photographs of October 7, 1959 possessed a surface resolution close to that of the early telescopes of Galileo Galilei, six years later—in July 1965—another Russian space probe, Zond 3, increased this resolution by a factor ten; and photographs such as reproduced on page 161 show surface details less than 2 miles in size—a limit which the American Orbiter 1 in August 1966 diminished to formations 90–270 yards across.

The techniques and accomplishments of the Orbiters will be more closely described in their turn in a later part of this chapter; but in this place we wish to add a few words on Zond 3 (which was probably a practice shot for Mars). As already stated, the latter was launched on July 18, 1965 into a heliocentric orbit; and its photographic mission during the lunar fly-by was accomplished on July 20, when the spacecraft approached the lunar surface at a distance varying between 5,700 and 6,100 miles. The optics of Zond 3 consisted of a telephoto lens of only 1.3 cm. free aperture and 10.6 cm. focal distance, operating at a focal ratio f/8 with exposures between 3 and 10 millisec. taken every 2¼ min.; and a total of twenty-five frames were obtained in the course of a 68 min. camera run. As for Luna 3, the films were processed aboard the space station, scanned in 1,100 lines per frame, and transmitted to the ground on July 29, when Zond 3 was already more than 1,400,000 miles from the Earth.

The principal contribution of the new evidence furnished by Zond 3—apart from the improved data on the topography of the Moon's far side in regions not previously covered by Luna 3—was the discovery, on the far side, of new types of surface formations which our Russian colleagues termed the "thalassoids"—large shallow depressions comparable in size with the smaller circular maria (like Mare Crisium or Serenitatis) on the near side, but lacking the darker material filling their floors. Such formations were subsequently discovered also on photographs taken by the American Orbiter 1 in August 1966 (see page 145) and their existence can be accepted without reservation. Therefore, it is not the mare basins which seem to be lacking on the far side, but rather the dark material (lava?) to fill them up; and although conjectures have been made concerning the reasons of this apparent disparity of the two lunar hemispheres we cannot as yet pretend to know the whole story.

Hard-landers

The hard-landing spacecraft launched to the Moon in the years 1959–65 fulfilled two essential purposes: to enable us to improve our knowledge of the mass of the Moon (essential for all subsequent operations of lunar astronautics), and to relay by television the close-up views of the lunar surface as seen from the spacecraft in the last

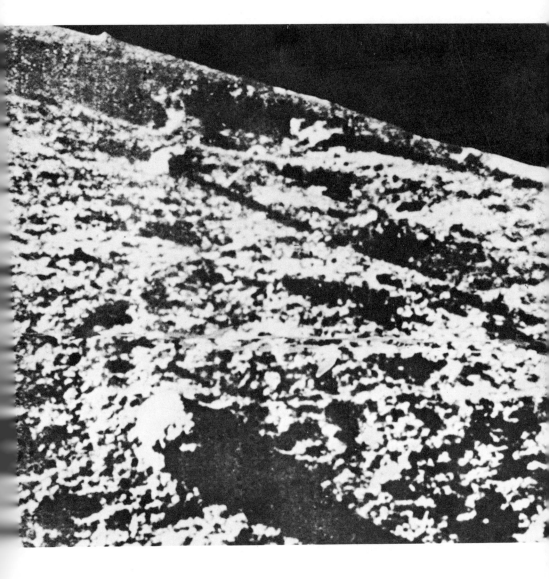

A view of the lunar landscape televised by the Russian soft-lander Luna 9 on 4 February 1966. (By courtesy of the USSR Academy of Science.)

quarter of an hour of its flight. While it is true that both these objectives had to await till 1964–5 for their accomplishment, let no one raise any regrets that this was not done at an earlier date; for he would be asking for the impossible.

In retrospect, it is entirely clear that the brilliant achievements of lunar missions during 1964–6 could have been attained only on the shoulders of their less-fortunate predecessors. The fact that the latter failed (to a varying degree) to fulfill all objectives of their missions represented a necessary process of learning by trial and error the intricacies of an entirely new branch of technology; and when the art was mastered the results more than justified all previous expenditures of effort. In a real sense it is true to say that the relative failures of the Rangers 3–5 were the price to pay for the triumphs of the subsequent mooncraft of this class and their successors; and today, it is clear to us that these triumphs were not bought at too high a price.

These introductory remarks should explain why, in what follows, we shall confine our attention mainly to the achievements of Rangers 6–9 in 1964–5. Even so, it is, however, impossible to by-pass without adequate mention some of the accomplishments of preceding spacecraft—such as the measurements of the lunar magnetic field by Luna 2 in September 1959 or of the γ-radioactivity of the lunar globe by Rangers 3 and 4 in 1962. The magnetometric measurements carried out aboard Luna 2 by a group of the Russian investigators (Dolginov, *et al.*) failed to detect the indications of any magnetic field above the lunar surface exceeding in strength 0.0003 gauss; and established that the mean magnetization of the lunar globe does not exceed 0.25 of a per cent of that of the Earth. Similarly, measurements of the γ-ray emission during the fly-by of the Moon by Ranger 4 prior to its crash on the far side revealed that the radioactivity of the lunar globe is not greater (and probably less) than that of the terrestrial crust—a result confirmed by the Russian soft-lander Luna 13 in December 1966.

The principal contribution to our science of the Rangers 7–9 in 1964–5 were, however, close-up records of the lunar surface which attained a resolution 1,000 times as large as that obtainable by the best ground-based telescopes so far; and revealed details a yard or less in size. To this end, the last block of this spacecraft carried a complement of six cameras in locations indicated by the photograph on page 191, three of which had 1-inch lenses working at a focal ratio f/1, three others 1.5-inch lenses of f = 75 mm. focal length. The former imaged a field 25° across; the other 8.4° across; in addition, others (P_1–P_4 cameras) provided smaller images 2.1°–6.3° in size. The A and B cameras (imaging a field of view of 25° and 8.4°, respectively) produced frames recorded by vidicon TV-tubes and scanned in 1,150 lines per picture; their exposures lasted but 2–5 millisec. (to arrest the motion of the spacecraft); but transmission (at 960 MHz.) occupied 2.56 sec. of real time (and reached the Earth with a time-lag of 1.28

sec.). The smaller images formed by the P-cameras were scanned in 300 lines; and their transmission took but 0.84 sec.; the energy required for transmission was 60 watts. The total weight of the spacecraft was 809 lb., of which optics and telemetry accounted for 379 lb.

The time of flight of the Rangers to the Moon was close to 65 hours; but their actual picture-taking mission occupied only the last quarter of an hour. The transmission commenced at an altitude of 1,300–1,550 miles above the lunar surface, and ended at 2,000–2,500 ft. above it; so that the remaining time of flight was sufficient to transmit only a part of the last P-frames. Altogether, Ranger 7 sent down 4,308 individual pictures; Ranger 8, 7,137; and Ranger 9, 5,814; so that their total contribution amounted to 17,259 individual frames.

What did this radically new evidence reveal to astronomers awaiting anxiously at the receiving end of the TV-transmission link? The nature of contributions so novel and on so massive a scale is difficult indeed to summarize in a few words; and any attempt to do so is bound to fall short of doing full justice to these remarkable experiments. With a ground resolution which exceeded that of

The Earth as seen from the Lunar Orbiter 1 on 23 August 1966 from a vantage point 1,000 km above the far side of the Moon. (By courtesy of the Boeing Company, Aerospace Division and of NASA.

the best previous terrestrial work by a factor of 1,000, it was possible to extend the lunar crater counts to formations three orders of magnitude smaller than any we had known thus far, and infinitely more numerous. Their statistics revealed for the first time the overwhelming abundance of secondary (and tertiary) craters, formed by impacts of, not primary cosmic intruders, but of boulders thrown out from other parts of the Moon by primary cosmic impacts. It revealed the clustering of such craters on the background of bright rays associated with some recent large primary impact craters (such as Copernicus or Tycho), as shown by the photograph on page 207 and disclosed for the first time the extent to which the lunar landscape must undergo local subsidence (probably triggered by moonquakes which must be produced by each major primary impact).

Close-up pictures of the interior of the crater Alphonsus, taken by Ranger 9, revealed the existence of maar-like formations of very low albedo (page 227) which are probably of internal (possibly volcanic) origin. The detailed structure of the ramparts of Alphonsus, revealing as it does (page 227) islands of floor-like ground in their midst, strongly suggests that these formations were not raised abruptly (by a catastrophic impact) but by a gradual tectonic process. Above all, the striking similarity in surface texture of the landing places of Rangers 7, 8 and 9—separated from each other (see page 89) by several hundred miles—strongly suggests that the processes which shaped it up were global rather than local in nature—i.e. external rather than internal in origin; for what else but external influences (i.e. sweeping up of interplanetary debris and dust) could impress the same nearly uniform type of surface relief all over the Moon?

One of the principal achievements in space exploration during the year 1966 was—not one—but three successful soft landings of man-made spacecraft on the lunar surface, and its investigation at a close range with the aid of the TV-cameras and other devices aboard such spacecraft. Three previous attempts (Luna 5, 6 and 8) were made by the Russians to soft-land a spacecraft on the Moon during 1965 before the final success of Luna 9 in the small hours (UT) of February 3, 1966—a success repeated by Luna 13 shortly before the end of that year. On the American side, the first attempt (Surveyor 1) to do so was brilliantly successful; but the second (Surveyor 2 on November 6, 1966) failed through equipment malfunctioning during the mid-course maneuver.

Before we give a brief account of the accomplishments of this novel type of mooncraft (see pages 119 & 123), let us say a few words about their equipment. At the commencement of their long journeys, the Lunas were very much heavier craft (325 lb. for Luna 5, 3,318 lb. for Luna 7, 3,422 lb. for Luna 8 and 3,840 lb. for Luna 9), than the Surveyor (2,194 lb.)—i.e. about three to five times as heavy as the Rangers—but the weight of the parts which soft-landed on the lunar surface were almost in an inverse proportion (596 lb. for Surveyor 1, in comparison with 220 lb. for Luna 9). Their flight schedules also differed to some extent; while the Russian mooncraft travelled in their 80-hour trajectories, the Surveyors accomplished their journeys in about 63 hours. More significantly, their trajectories differed also in their terminal decelerations. For Surveyor 1 this began at an altitude of 52 miles above the lunar surface, when the mooncraft's velocity was 8,515 ft./sec. At an altitude of 5.3 miles this was decelerated to 396.5 ft./sec.; and the touchdown speed was only 9.75 ft./sec.—in comparison with about 32.5 ft./sec. for Luna 9. The terminal deceleration of the latter, while "soft" for the instrumental package aboard, would probably have been lethal for man; while the gentler deceleration of the Surveyor would have given him a good chance of survival.

The Surveyor mooncraft which landed on the Moon on June 2, 1966, 6 hr. 17 min. 36 sec. UT—12 ft. across and 10 ft. in height—was physically much larger than Luna 9 or 13, the dimensions of which were only 5 ft. across and 3½ ft. in height. After landing, the Surveyor was powered by solar panels whose silicon cells provided electrical power of 85 watts, only a small part of which (about 10 watts) was used for transmission to Earth at a frequency of 2,295 MHz. The Lunas were powered by chemical batteries, giving them a useful lifetime of only three to five days (while Surveyor 1 functioned almost as many lunar days) and transmitted at a lower frequency of 183.53 MHz.

Little is known of the optics or TV systems of the

Lunas; but those of the Surveyor were similar to those employed by the Rangers. Their optics consisted of a dioptric system of 1-inch free aperture and variable focal length which ranged from 25 mm. (giving a field of view of 25.4° x 25.4°) to 100 mm. (field 6.4° x 6.4°); the exposure times being of the order of 0.1 sec. A periscopic arrangement made it possible for the Surveyor to scan the surrounding landscape from a minimum distance of 47 in. to the horizon about a mile away;* and while Luna 9 or 13 could inspect lunar ground at a similar minimum range, the fact that its lenses were only 22 in. above the ground gave them a much smaller maximum range of vision. The TV cameras of the Surveyor were vidicon tubes similar to those employed by the Ranger, with 200–600 lines per picture; and for Luna 9 the number of lines per frame was likewise close to 600.

With this equipment, Surveyor 1 related to Earth more than 10,000 individual frames over the period of its activity which extended well into the autumn of 1966. What kind of ground did its TV cameras reveal during their horizontal periscopic sweeps? Like Luna 9 on February 3 or Luna 13 on December 24, Surveyor 1 landed on a dark and gently rolling relatively smooth surface (nearly level on a kilometer scale) of a typical mare ground, sparsely dotted with boulders of various size and shape, and small craters (from several inches to

---

* It is well to keep in mind that, on a spherical globe of the size of the Moon, a camera situated 63 in. above the ground would encounter the horizon at a distance of 1.6 mile.

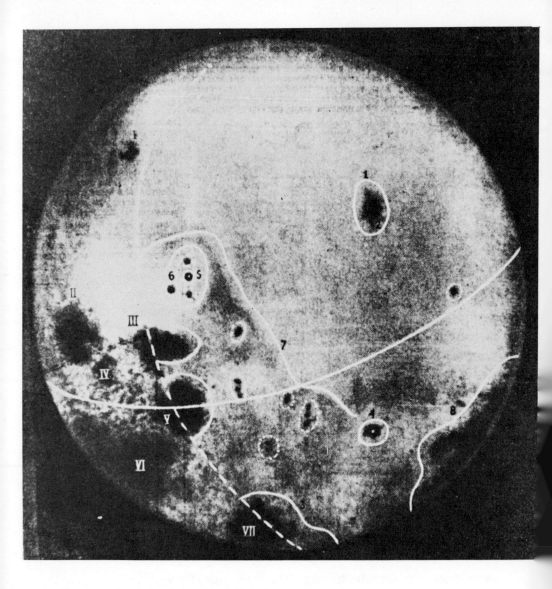

A photograph of the far side of the Moon, secured by the Russian Luna 3 on 7 October 1959, at a distance of 66,000 km. The solid arc marks the position of the lunar equator, the dotted arc, the limit of visibility from the Earth.

many yards in dimensions) which are obviously of secondary impact origin. The largest of such craters, visible in outline on the horizon, is almost a half mile in size; and the outlines of a few low mountains protrude from below the horizon at a distance of a few miles.

In the immediate neighborhood of the spacecraft, the lunar surface is seen to consist of a granular material of a rather wide size range: coarser blocks of rock and smaller fragments are interspersed with particles a millimeter or less in size. This material was disturbed and penetrated by the spacecraft's footpads to a depth of a few inches (see page 175). The nature of this disturbance reveals that the lunar surface material possesses a definite amount of cohesion and does not consist of loose dust. On its final approach to the lunar surface, the Surveyor's vernier retro jet engine functioned till an altitude of a mere 13 ft. (at which the spacecraft was decelerated to a velocity of 1 sec., which increased again at touchdown to 9.75 ft./sec. by free fall in the lunar gravity field); and no surface material appears to have been dislocated by their action to any appreciable extent.

In order to test further the cohesion of lunar material, the spacecraft's altitude control jet (of nitrogen gas) on one of its legs was operated repeatedly some 6 in. above the surface, and its action observed by the Surveyor's TV cameras at a close range—but the pictures transmitted to the Earth failed to indicate any surface disturbance. In addition, the thermal measurements made aboard the Surveyor indicated that the mooncraft's temperature control surfaces were not covered by any dust that could have been stirred at the time of the landing. From the extent to which the lunar surface yielded to the Surveyor's footpads (and to which these footpads were compressed) on touchdown, the lunar ground appears to possess a static bearing capacity of about 5 lb. per sq. in. (i.e. 3×10^5 dynes/cm^2)—a result consistent in its order of magnitude with the measurements performed by Luna 9 and 13 in different parts of the mare ground. This corresponds to a bearing strength of (say) wet sand, and is sufficient to support the weight of a man carefully making his way on such a surface.

The pictures transmitted to the Earth by Luna 9 (one of which is reproduced on page 141) were taken at the time when the altitude of the Sun rising above the lunar horizon was between 7 and 41 degrees. The much longer operating period of Surveyor 1 on the Moon gave it an opportunity to record for us the view of the lunar landscape at sunset—with the Sun much closer to the horizon —bringing vividly home to us the utter desolation of the lunar environment (see page 112). The fact that the Surveyor's cameras could also function some time after sunset (before the ambient temperature became too low) made it possible for them to relate to us the view of the solar coronal streamers protruding above the horizon in airless space—an example of which is shown in the photograph on page 55. Stars well below the limit of visibility of the naked eye were located on views directed to the sky, indicating the glory of the heavens that one day will be seen by man from the lunar vantage-point.

Surveyor 3—the next member of the American soft-landing family to accomplish its purpose—added to previous instrumentation a remotely controlled surface sampler (see page 192)—a device designed to test the hardness and texture of the lunar surface material in the view of the television camera—tasks which Luna 13 attempted to accomplish by different technical means. Moreover, Surveyors 5–7 were provided with (in addition to the television camera and a mechanical surface sampler, a chemical sampler of the atomic composition of the lunar surface—in the form of the α-scattering experiment box shown on page 236. The sensor unit in the box contains a radioactive source (curium 252) of α-particles which bombard the section of lunar surface immediately below the experimental box; and also a detector of back-scattered α-particles and protons, from the distribution of which the atomic composition of the irradiated surface can be inferred. A successful repetition of this experiment in the three different landing places of Surveyors 5–7—two in the lunar maria and one in a continental region (close to the crater Tycho) disclosed that the atomic composition of the lunar surface appears to be close to that of basaltic rocks well known on the Earth, whose density (3.2–3.3 g/cm^3) comes also very close to that of the lunar globe.

It is, therefore, time to say with Homer Newell that while Surveyor 1 put man's eyes on the Moon, Surveyor

3 added an arm and a hand with which to work; to which Surveyor 5 may have added the sense of taste. Moreover, just as Surveyor 1, not content to scan the lunar panorama alone, turned its television eyes to the solar corona and the stars—Surveyor 3 happened to be the first soft-lander to witness (on April 24, 1967) the eclipse of the Sun by the Earth on the lunar surface (an event subsequently experienced by Surveyor 5 on October 18 of the same year), when for 41 minutes of total phase sunlight was completely cut off. During the entire eclipse lasting 107 minutes, Surveyor 3 experienced, and reported to Earth, precipitous temperature changes amounting to almost 200° C; thus confirming the results of previous work performed from our terrestrial home bases. However, its cameras were the first to relay to us the view from space of the terrestrial aureola surrounding the rim of our planet on such occasions (due to diffraction of sunlight in our atmosphere). Altogether, the five successful Surveyors relayed to the Earth over 80,000 television pictures of the lunar surface as seen from their respective vantage points—almost as many photographs as ever taken of the Moon from all terrestrial observatories—of which only a few can be reproduced in this book; (see pages 55, 112, 203); and Surveyor 7 had, in addition, the crowning glory of recording for us man-made flashes of laser light beamed to the Moon from the United States—as forerunners of optical messages which will one day bear the brunt of communications between our Earth and its satellite.

As to Luna 13—the last Russian soft-lander in 1966 —which came to rest again in the plains of the Oceanus Procellarum near the crater Seleucus, its 220 lb. capsule contained a television system similar to that aboard Luna 9; but the fact that the capsule landed on the slopes of a small crater restricted its field of view to the immediate neighborhood of the spacecraft. In addition, the latter carried a soil density meter (hammerhead with titanium point) driven into the lunar soil by a jet device mounted on the end of one experimental boom. A second density gauge (γ-radiation source, coupled with a gas discharge counter to measure the flux of dispersed radiation) was mounted on the other boom, and furnished significant results.

Orbiters

The second signal achievement in space research of the solar system during the years 1966–68 was the launching and injection into circumlunar orbits of not less than ten artificial satellites—four from Russia and six from the United States—three of which are still circling the Moon today. Their designations and the dates of their launch have already been listed in Table II (d); but before reviewing their accomplishments a few words may again be in place concerning their equipment.

Little is, unfortunately, known so far on the instrumental packages aboard the Russian satellites Lunas 10-12 or 14; but that of the American Orbiters (see page 211) is well known. The principal mission of the six satellites of this family (two of which have already been successfully launched) is to photograph with moderate and high resolution selected parts of the lunar surface on which manned landings may be attempted in the next few years; and to this end this mooncraft has been provided with two photographic cameras: the medium-resolution camera had an 80 mm. focus Xenotar lens working at f/5.6; while for high-resolution work a 600 mm. focus Paxoramic lens was used at the same focal ratio. The image was not formed, as on the Rangers and the Surveyors, in a TV vidicon tube, but (as for the Russian Luna 3 or Zond 3) on 70 mm. photographic film, whose 200 ft. roll is sufficient to record 211 frames with each lens. The exposure times ranged between 10 and 40 millisec. The films were developed aboard, scanned electronically and the results telemetered to the Earth, where the image is reconstituted in strips. The transmission of each full frame (70 mm. in width), consisting of one medium-resolution and one high-resolution picture, required about 43 min.—so that during each revolution (of approximately 200 min. duration) the Orbiter could transmit only about two frames (while the spacecraft was visible from the Earth).

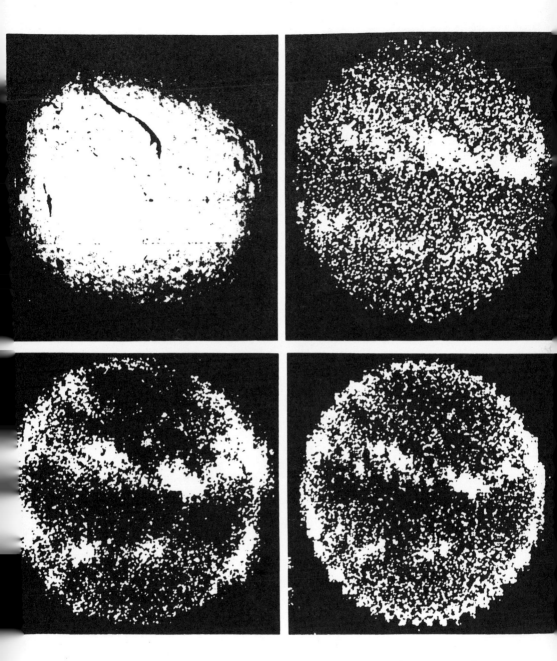

Monochromatic images of the Sun digitally transmitted by OSO-4 between 25-27 October 1967 in the light of the O VI line at 1032A (above, right), of Lyman continuum at 897 Å (lower left) and of the Mg X line at 625Å (lower right). In the upper left, a conventional Hx spectroheliogram taken at the same time. (By courtesy of NASA.)

The primary missions of both Orbiters were to photograph in detail thirteen sites provisionally selected as possibilities for future manned landings on the Moon. In order to map them, overlapping pictures from the moderate-resolution camera are being used as stereo pairs, to be reduced by stereogrammetric techniques. The high-resolution coverage is not stereoscopic, and such photographs are being reduced individually by standard photometric methods.

Approximately 90 per cent of the entire film supply was used to map the prospective landing sites on the Moon for NASA; while the balance was used up by scientists of the Boeing Company (prime contractor for the Orbiter project) for "housekeeping purposes" to check the performance of the individual parts of the equipment. Photographs so obtained of the far side of the Moon by Orbiter 1 (see page 145) or of the oblique views of the crater Copernicus by Orbiter 2 (see pages 205 & 212) belong among the most interesting results of their missions, and have rightly been referred to as "pictures of the century" by the scientific public; and the same can be said of many photographs taken by subsequent spacecraft of this class (pages 145, 170, 205, 212)

The U. S. Orbiter mooncraft (see page 211) are smaller than the Surveyors (about 5 ft. in height and width with panels folded, but 12 ft. wide across extended solar panels, and 18½ ft. across extended antenna booms), and weighing but 850 lb. are considerably lighter than the Russian Lunas 11–12 weighing 3,600–3,700 lb. Their

extended solar panels (consisting of over 10,000 individual cells) delivered electrical energy of 375 watts;
in addition, a 20-cell nickel-cadmium battery supplied a
current of 22–31 volts.

The flight plan for Orbiter 1 called for a cautious
approach to its target. Following its launch on August
10, 1966, and its injection in lunar orbit three days later,
it spent almost a week in a "parking orbit" at a mean
altitude of about 600 miles above the lunar surface as
a sensor of the lunar gravitational field, on which very
little direct information was available up to that time.
Only when a preliminary analysis of accurate Doppler
tracking revealed the magnitude of its principal harmonic
terms was the spacecraft lowered cautiously into its
"working orbit" for photographic mission, which at the
time of the periselenium passage approached the lunar
surface within approximately 40 miles. It remained in
this orbit (the elements of which continued slowly to
vary as a result of the quadrupole and higher terms in
the gravitational potential of the lunar globe) and re-
sponded faultlessly to hundreds of commands from the
Earth, including the last one which hurled it on October
29 to its destruction on the Moon, in order not to interfere with the signals of its successors.

Orbiter 2 was launched on November 6, 1966, and
injected into circumlunar parking orbit on the 10th,
in which it spent the next five days to serve (like Orbiter
1 before) as a passive probe of lunar gravitational field.
Its descent into a more eccentric working orbit (similar

to that of Orbiter 1) commenced on November 15; and the period of its photographic mission continued from November 18 until December 6. Orbiter 3 followed in February 1967, Orbiter 4 at the beginning of May; and Orbiter 5—the last member of this family and scientifically by far the most revealing one—in August. The planned destruction of Orbiter 5 by a crash on the lunar surface, which occurred on January 29, 1968, brought to a close one of the most successful episodes in the exploration of the Moon by spacecraft—when within a time-span of less than two years we learned to know the topography of both sides of the Moon almost as well as we know it from our own Earth.

Turning to other types of spacecraft, little is known of the mission objectives of the Russian Lunas 10–12 or 14 so far. The 3,616-lb. instrumental package of Luna 12, launched on October 22, 1966, reportedly contained photographic equipment in its nearly equatorial orbit at a distance from the lunar surface varying between 61–1,081 miles; but if so, no results of its work have so far been made public. An analysis of the lunar gravitational field from secular perturbations of Luna 10 has, however, been published (Akim); and a comparison of his results with those of the early Orbiters (Michael, et al.) ·by Goudas, Kopal and Kopal disclosed an appreciable correlation between external gravitational potential and surface deformations—indicating the Moon to be a largely homogeneous globe.

Explorer 35—a 230-lb. lunar-anchored IMP, which

A more detailed view of a part of the far side of the Moon, secured by the Russian spacecraft Zond 3 on 20 July 1965. (The bright disc in the lower left-hand corner is a phtometric scale.)

since July 22, 1967 has been revolving around the Moon in a 684-hour eccentric orbit at a distance varying between 500–4,600 miles from the lunar surface—represents the smallest artificial satellite sent out by human hand to circle around the Moon so far. As it carries, however, no optics, its activities do not fall properly within the scope of this book—beyond a mere mention that its mission is essentially electromagnetic. From its vantage point in space, this spacecraft measures the Earth's magnetosphere and magnetic tail as it engulfs the Moon every synodic month. It measures also the effects of interaction between the lunar globe and the solar wind; and an analysis of such measurements has already indicated that the strength of the magnetic field of the Moon—if any—must be less than 10^{-6} gauss (i.e., much less than the limit indicated by Luna 2 in 1959). In the light of these results the Moon appears, therefore, to be essentially a solid rock devoid of magnetism, and internally cold.

Planetary Probes

Since the first man-made spacecraft disengaged themselves from the gravitational field of our mother-planet, the Moon did not long remain their sole target; and up to the time of writing not less than eleven spacecraft of American as well as Russian origin ventured farther afield—to pay close calls on our next two celestial neighbors in space: namely, the planets Venus and Mars.

A list of these spacecraft (limited only to those which

attained heliocentric orbits) is given in the accompanying Table III. Only six of them—three American and
three Russian—reached the close proximity of their astronomical targets: namely, the U. S. Mariner 2, 5 and
U.S.S.R. Venus 2, 3, 4 for Venus and Mariner 4 for
Mars. Mariner 2—the first really successful interplanetary flight—by-passed Venus on December 14, 1962
(i.e. 109 days after its launch) at a minimum distance
of 21,594 miles from the planet's center, and continued
to furnish data on interplanetary medium *en route* until
the first half of January 1963, to a distance of 54 million miles from the Earth, after the probe had travelled
more than 225 million miles along its mildly eccentric
346-day orbit (inclined 1.2/3 degrees to the ecliptic)
around the Sun.

The Mariner spacecraft (see page 230) weighs 447 lb.
and, in the launch position is 5 ft. in diameter and 9 ft.
11 inches in height. In the cruising position with solar
panels and antenna extended (as on page 230) it measures 16 ft. 6 inches across and 11 ft. 11 inches high. Its
hexagonal framework houses a liquid-fuel rocket motor
for trajectory correction, and six modules, containing
the altitude control system, electronic circuitry for the
scientific experiments, power supply, etc. Equipment for
the experiments themselves is attached to a tubular
superstructure extending upward from the hexagonal
base. An omnidirectional antenna is mounted at the
peak of the superstructure (see page 230), and a
parabolic high-gain antenna is located below the base

of the hexagon. Two solar panels, 27 sq. ft. in area, consisting of 9,800 silicon cells, delivered electrical power ranging between 148 and 222 watts at different distances from the Sun. A command antenna for receiving transmissions from the Earth is mounted on one of the panels.

The Mariner 2 spacecraft contained six scientific experiments, two of which are of primary interest for telescopic astronomy: namely, a microwave radiometer capable of scanning (albeit with low angular resolution) the image of Venus at wavelengths of 13.5 and 19 mm. (i.e. radiation which penetrates the Cytherean cloud veil); and an infrared radiometer scanning at wavelengths between 8 and 11µ radiation emitted by the top of the cloud layer. In addition, a magnetometer was carried aboard to measure the strength and direction of any Cytherean as well as interplanetary magnetic field; and an ionization chamber with three Geiger counters to measure the number and intensity of energetic particles (primary cosmic rays) in interplanetary space.

The tail end of the 40-inch Yerkes refractor, with a nebul
refractor attached to its side.

The results obtained with the aid of this equipment in the course of the spacecraft's 130 days of experimental lifetime covered an impressive range of diverse phenomena. During its rendezvous with Venus accurate two-way Doppler tracking of the spacecraft's motion furnished a much-improved value for the mass of this planet. The latter was notoriously difficult to obtain otherwise, as Venus possesses no natural satellite. However, the measured reflection of the spacecraft's trajectory in the proximity of Venus, caused by the latter's gravitational attraction, disclosed the mass of Venus to be equal to 0.81485 ±0.00015 times that of the Earth.

The results obtained with the aid of the two radiometers aboard were no less valuable. At the time of the closest approach (between 2 hrs. 59 min. and 3 hrs. 34 min. p.m., EST, December 14, 1962) both instruments scanned Venus (then visible from the spacecraft as an apparent disc of 21.4° angular diameter illuminated at "first quarter") three times—first on the dark side, then along the terminator line, and finally on the illuminated side; and a total of eighteen data readings were made during the scans. Microwave scans at 13.5 mm. wavelengths included atmospheric as well as surface radiation (though predominantly the former if the atmosphere were a strong microwave absorber—such as it would be if it contained an appreciable amount of water). On the other hand, the 19 mm. wavelength is not affected by water vapor; and radiation received at it probably refers to the planet's surface. Thus the larger the brightness

temperatures recorded by the two channels of the microwave radiometer, the greater the amount of water vapor
in the atmosphere of the planet.

The effective temperature obtained by the microwave
radiometer were close to 450°K (180°C) on the night
side, 550°K (280°C) near the terminator (i.e. line of
sunrise), and only 390°K (110°C) in daylight. These
figures disclosed strong effects of limb-darkening; but
no evidence was obtained for the presence of any water
vapor.

The infrared radiometers operated at two wavelength
ranges of 8–9μ and 10–11μ, selected so because the absorption by carbon dioxide could strongly affect the latter
but not the former. Any marked difference in signal intensity would indicate a break in the cloud layer surrounding the veiled planet, as it would allow the shorter
but not the longer wavelength band to penetrate more
deeply towards the surface. In actual fact, both channels
indicated virtually the same temperature (close to 235°K
or —38°C) which clearly refers to the top of the cloud
layer visible from the Earth. No breaks in the clouds
were seen from the vantage-point of the spacecraft; and
a scan carried out at the same time with the 200-inch
reflector of Palomar Mountain, using a radiometer operating in the atmospheric transmission window between
8 and 12μ's, gave results in good agreement with the
measurements carried out in space.

In addition, the magnetometer aboard Mariner 2
could find no evidence of any Cytherean magnetic field

at any point along the spacecraft's trajectory. Comparisons of Mariner 2 magnetic observations in the vicinity of Venus with the Pioneer 5 measurements near the Earth indicate that the strength of the magnetic dipole moment of Venus, if any, is less than 0.05–0.10 of that of the Earth—perhaps not an unexpected result, in view of the slowness of the axial rotation of Venus; and in line with the vanishing magnetic field of the Moon.

Perhaps the most valuable set of the data provided by Mariner 2 in the course of its 130-day career as an instrumented probe was the extensive evidence on the particulate contents of interplanetary space and its magnetic fields. These subjects fall, however, largely outside the scope of this book concerned primarily with optical astronomy; and for their more detailed survey the reader must refer to other sources. Owing to low resolution generally attainable in the infrared and microwave domain, the contributions of Mariner 2 to optical astronomy were, perhaps, not too extensive; and neither were those of the Russian space probes Venus 2 and 3 which followed Mariner 2 on the same journey three years later. Venus 2, launched on November 12, 1965, missed on February 27, 1966, the planet by only 15,000 miles—i.e. approached it by 6,000 miles more closely than Mariner 2 did in 1962. Two days later Venus 3 actually crashed on the Cytherean surface and thus became the first man-made spacecraft to land on another planet. Throughout the entire flight regular radio communications were maintained with the spacecraft, and scientific information on interplanetary environment received *en route*. However, communications are said to have failed in the last twenty-four hours before approach; and whether or not Venus 3 did anything else than to crash-land on the planet was not revealed.

The 1967 conjunction between the Earth and Venus attracted to the latter two further spacecraft sent out to carry out further exploration of the veiled planet: namely, the U.S.S.R. Venus 4, and the American Mariner 5; and both rendezvous met with a large measure of success.

Venus 4—a 2,438-lb. spacecraft launched on June 12, 1967—successfully reached its destination on October 18 after a journey of 128.4 days through interplanetary space; and released an egg-shell capsule about 1 yard in size, coated with an ablative heat shield that burned away as the capsule entered the Cytherean atmosphere and plunged into the clouds. Once it had been sufficiently slowed down by air drag, the capsule released a parachute and, with its aid, completed its descent in 94 minutes—thus accomplishing the first soft landing on any planet.

The measurement carried out by the instrumental package en route indicated that the atmosphere of Venus consists of about 98 per cent of carbon dioxide (CO_2); the rest being water vapor or molecular oxygen. Surprisingly, no trace of nitrogen was found in the records. The temperature prevailing at some 16 miles above the surface was reported to be about $313°K$ ($+40°C$), but rose to $553°K$ ($+280°C$) at the end, where air pressure be-

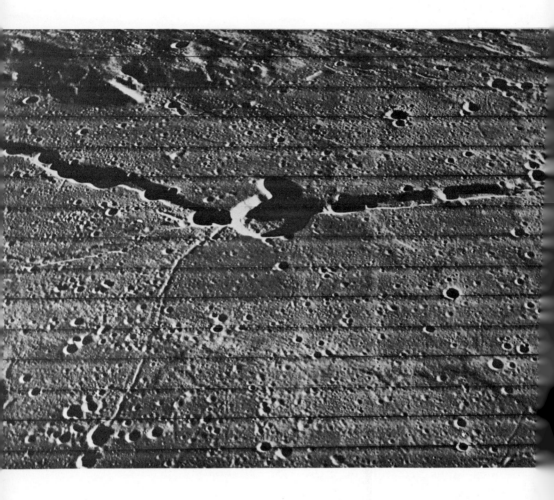

The well-known lunar Hyginus rille (between Sinus Medii and Mare Vaporum) from a new angle — as photographed by the Lunar Orbiter 3 in February 1967 from an altitude close to 50 km above the lunar surface. (By courtesy of the Boeing Company and of NASA.)

came 15–22 terrestrial atmospheres. It is not certain whether these final readings were made on the planet's actual surface; for intense heat may have destroyed the electronic recording equipment in the capsule before it reached the ground.

The American Mariner 5—a substantially lighter spacecraft (540 lb.)—commenced its four-month voyage to Venus on June 14, 1967; and on October 19—1.5 days after the Russian touchdown, Mariner 5 by-passed Venus at a minimum distance of 2,480 miles from the surface, while its instruments viewed the planet and sampled its environment. Mariner 5 sent also radio-signals (at 10-cm. wavelength) through the Cytherean atmosphere as it was occulted by the planet, and reappeared 21 minutes later to bid farewell to Venus and depart along a heliocentric orbit to become a new asteroid of the Sun. Telemetry of all data back to the Earth across 49 million miles with the available power of 10 watts took 34 hours of real time.

A preliminary analysis of these data led to a number of significant results. Thus Mariner 5 detected no belt of charged particles (similar to the terrestrial van Allen belts) girdling Venus. The data are consistent with at most a weak planetary magnetic field of strength not more than about 0.01 gauss (the Russian probe likewise failed to detect any Cytherean magnetic field at all). According to Mariner 5, the atmosphere of Venus at the height of its closest approach consists of 75-85 per cent carbon dioxide, with some traces of hydrogen (but not oxygen).

Turning our attention to Mars—our nearest outer celestial neighbor—the situation for a long time was no better. The Russian Mars 1 probe—the first one known to be sent out to the red planet—made valuable contributions to our knowledge of the particular contents of interplanetary space during the first 66 million miles of its flight; but none to optical astronomy. With its successor—Mariner 4 to Mars—it was a very different story.

Mariner 4 was launched from Cape Kennedy in the United States on November 28, 1964 to embark on its seven-month journey through space to its rendezvous with Mars on July 14 of the next year, and accomplished its mission with brilliant success: for at the time of its closest passage it overflew the Martian surface at a height of a mere 6,118 miles, after a journey of 140 million miles—a marksmanship which may be likened to rolling a strike in a bowling lane at a distance of 400 miles! And not only that; but after so long a journey all instruments aboard the spacecraft wakened up to all radio commands from the Earth, which took 12 min. and 30 sec. to reach it, and performed faultlessly a mission which can be truly termed historic.

The principal equipment aboard—as far as we are concerned—was a camera the optical system of which consisted of a Cassegrain reflecting telescope of 4.5 cm. free aperture and 36 cm. focal length, operating at an equivalent focal ratio of f/8. The receptor was a miniaturized all-electrostatic vidicon tube, whose output (the target current) was amplified and digitized into $2^6 = 64$ "quantum levels" of intensity. Each frame recorded by this equipment consisted of 200 lines of 200 picture points, the scans of which furnished 200 x 200 x 64 == 256,000 bits (i.e. pulses which were present or absent) of information that had to be transmitted to the Earth. This transmission was accomplished at a frequency of 960 MHz. with an energy of a mere 10 watts—which by the time it reached the Earth was reduced to one quintillionth (10^{-18}) of a watt—a very small fraction of some 10^5 watts at which a typical television station on Earth transmits its video signals (picture). One price which had to be paid for this tremendous disparity in power was the speed with which pictures could be transmitted to us from the distance of Mars. The latter were recorded by an all-electrostatic vidicon tube aboard the spacecraft with exposure times of only one-fifth of a second; but since (for the bandwidth employed) the bits of information of which these images consisted could be sent out at a rate of not more than 8.3 per sec., it took more than eight hours of real time to transmit more than a quarter of a million bits of data constituting each single frame.

In this manner, a total of twenty-two views (twenty-one complete, and a part of another) of the Martian surface were transmitted to us during the days following Mariner 4's historic fly-by of Mars on July 14–15, 1965. They were taken from altitudes ranging from 10,500 to 7,400 miles above the Martian surface (at the time of the minimum approach of 6,118 miles, Mariner 4 was on the night side of Mars); and their ground resolution —corresponding to a separation between individual lines of the televised picture—was approximately 2 miles. The latter corresponds to an angular resolution of 0.007″, exceeding that attained by our best telescopes from the distance of the Earth at least a hundred times. This is what we gain by sending out telescopes—albeit small— to close proximity of the object!

What did the surface of Mars look like from the vantage-point of our spacecraft? A glance at the photographs obtained in the course of this historic experiment—a sample of which is reproduced on page 109—reveals an arid landscape, unobscured by any kind of clouds, and profusely dotted by pockmarks which represent so characteristic a feature of the surface of any solid celestial body unprotected by an adequate atmosphere: namely, the impact craters. These are indeed so familiar to any casual observer of the Moon through the smallest telescope as to need scarcely any further identification in this place. That the surface of Mars (only surmounted by a very tenuous atmosphere, offering but little protection from impinging meteorites) should look essentially similar to that of the Moon could indeed have been anticipated; but still the outcome of the Mariner 4 television experiment came out as a distinct surprise; and was eagerly seized upon by the areologists and other students of the solar system to advance our knowledge of some of the processes which have been operative within it since the days of its formation.

One of the footpads of the Surveyor, which penetrated a few inches into the lunar ground at touchdown. (By courtesy of the JPL, CIT.)

Although the pictures secured by Mariner 4 cover only a little more than 1 per cent of the entire Martian surface, they show about seventy craters with diameters between three and seventy-five miles; and if that part of the planet surveyed by Mariner 4 can be taken as representative of the Martian surface as a whole, the entire planet may be pitted by more than ten thousand similar formations —not to speak of smaller craters whose numbers may be (like on the Moon) beyond counting.

Are these numbers too small, or too large, in comparison with those of the impact craters on the Moon (or on Earth)? For a given density of meteoric (or cometary) bodies in space, the cumulative number of hits scored at random at any solid target depends only on the duration of this kind of celestial target practice— counted from a time from which the oldest landmarks are still preserved on the respective score-board; and in the face of the crater counts on Mars and the Moon, the question can be asked: Are the oldest landmarks on the surfaces of both these bodies of comparable age? If the oldest craters on Mars were as old as those on the Moon (where their age goes back virtually to the origin of the Moon as a celestial body), this could mean that—like the Moon—Mars never had significant amount of water above its surface, or an atmosphere much denser than the present one; for otherwise features which are areologically old would have been—like on the Earth—gradually obliterated by erosion and denudation. If, on the other hand, the craters on Mars are deficient in number as compared with the density of similar formations per unit area of the lunar surface, this would imply that—at least in the past—the Martian surface must have been swept by wind, water or other erosive forces sufficient to level off the earlier landmarks.

At the present time, no definitive answer to this question can as yet be given. However, the crater counts based on the Mariner data—extending, to be sure, over a little more than 1 per cent of the entire Martian surface —indicate a crater density per unit area on Mars to be less than that encountered in continental areas of the Moon, though greater than in the lunar maria. Had both these celestial bodies been exposed to an equal flux of meteoritic bombardment from space, it would follow that the oldest landmarks visible now on Mars are younger than those on the Moon—indicating that air and water may have played a greater role in shaping up the sculpture of the Martian surface in the earlier stages of the history of the red planet than has been the case in the more recent past (i.e. posterior to the formation of the lunar maria); but we cannot as yet be sure.

A testimony against an excessive role of water at any time of the Martian past is the absence, on Mariner photographs, of any obvious features that could correspond to the basins of former oceans, or fossil river beds. That a small amount of water is present still now in the Martian environment is, of course, attested by the seasonal changes in Martian polar caps, and the appearance of water-vapor absorption in the spectrum of the planet.

The first view of the Earth from Space — a composite photograph of the South-western part of the United States obtained by V-2 rockets from 100 miles above.

Mariner 4 photographs—such as the one reproduced on page 109—revealed evidence of what may well be hoarfrost around the rims of some Martian craters at elevated altitudes (above 9,750 ft.)—a hoar-frost which may turn directly to vapor and back again to frost without becoming liquid in between.*

These revelations on the topography of the Martian landscape at a close range represented probably the most important contributions to our knowledge of our nearest outer planetary neighbor; but by no means the only ones of lasting significance. Thus the refraction of the radio signals sent out by Mariner 4 on its occultation by Mars disclosed that the density of the Martian atmosphere is much less than had previously been surmised from indirect terrestrial evidence; and that the air pressure prevailing above the Martian surface is only between 8 and 10 millibars (i.e. about 1 per cent of the terrestrial air pressure at sea level—met in our atmosphere at an altitude of some 19 miles); moreover, its mass appears to consist predominantly of carbon dioxide (CO_2).

Furthermore—like the Moon or Venus—Mars does not also possess any measurable magnetic field; and in order to escape detection the magnetic moment of Mars must be less than 2–3×10^{-4} of that of the Earth. Unlike Venus, Mars is rotating fairly fast about its axis (one day on Mars lasting 24 hrs., 37 min. and 23 sec. of our own time); but its mass (0.108 of that of the Earth)

* Unless, perchance, these white patches are due to solid carbon dioxide.

is probably too small to have generated a metallic core. In consequence, it did not come as a surprise to learn from the particle counters aboard Mariner 4 that, like Venus, Mars is not surrounded by any "van Allen belts" of charged particles; though it may possess an ionosphere dense enough to reflect radio waves transmitted from its surface in daytime.

What Mariner 4 did *not* find was any indication of life on Mars (nor any trace of the so-called "canals"); but it was not intended to do so. Its mission blazed, however, the way for later spacecraft to land instruments— and, eventually, men—on Mars; and thus to extend the domain of experimental astronomy to this nearest of our planetary neighbors. The U. S. National Aeronautics and Space Administration already plans a new mission—the Voyager—to land an instrumented package on the Martian surface in the early 1970s; and—all going well—to effect manned landings around 1985. Much remains yet to be done before these great goals will have become accomplished facts; but it is just possible that, within our lifetime, we may get in touch on the Martian surface— for the first (and last) time in the history of the solar system—with indigenous living matter which originated quite differently from the life as we know it on Earth. Isn't this prospect really worth while living for until 1985?

Concluding Remarks

The impact on human science of recent advances in lunar and planetary studies made possible by deep-space probes operating in close proximity to their targets, summarized briefly in this chapter, cannot be over-estimated. It is, in fact, true to say that greater progress has been made in this field in the past seven years than in all the preceding decades and centuries since the discovery of the telescope. In 1959, the first photographs of the Moon's far side were obtained from spacecraft in orbit. In 1964–5, the first closeups of the lunar surface were taken in three distinct localities by the Ranger spacecraft which (within the areas covered by the last frames of each Ranger) resolved details less than a yard in size—thus exceeding the best resolution obtainable from the Earth by a factor of 1000. Moreover, since 1966 seven lunar soft landings have been effected, which in the immediate neighborhood of the spacecraft increased this spatial resolution by a further factor of 1000—thus revealing on the Moon surface details less than a millimeter in size—and performed many other measurements which would have been utterly impossible at a distance. In addition, five U. S. lunar orbiting satellites in 1966-67 provided a photographic coverage of the entire globe of the Moon with a resolution of some 270 yards, and of about 38,600 sq. mi. of the lunar surface with a resolution of 1–2 yards. On the planetary side, the accomplishments of Mariner 2 in 1962, Venus 2 and 3 in 1966 and Mariner 5 with Venus 4 in 1967 to Venus, and Mariner 4 in 1965 to Mars, all made history in planetary studies; and revealed to us for the first time the real nature of the Cytherean atmosphere and of the Martian surface.

The optics carried aboard all these spacecraft were very modest in size, judged by the standards of contemporary ground-based astronomy: the largest objective launched so far into space—the 4-in. Paxoramic lens of the high-resolution camera of the U. S. Orbiters—is nothing much to speak of in comparison with the standard equipment of ground-based astronomical observatories. However, it was the ability of our spacecraft to operate their equipment in close proximity of their celestial targets which made their contributions so overwhelming; and there is but little room for doubt that future exploration of the solar system will depend to an increasing extent on this strategy.

What we learned so far is, to be sure, impressive enough; and many ghosts haunting previously our field were laid to rest by their action. Thus we know now that the lunar surface is not covered by any thick layer of loose dust, or "fairy castles"; that there is no excessive radioactivity on the Moon; nor are there "canals" on Mars. Instead the surface of the Moon appears to be reasonably hard (with a static bearing strength of the order of a few pounds per square inch) and very rough on a cm.-mm. scale—as we suspected long before from the observed lunar light changes and radar echoes. Moreover, the surface of Mars appears to look very much the same.

Secondly, none of the new evidence furnished by the Rangers, Lunas, Mariners, Orbiters or the Surveyors (albeit limited so far to only a small fraction of the

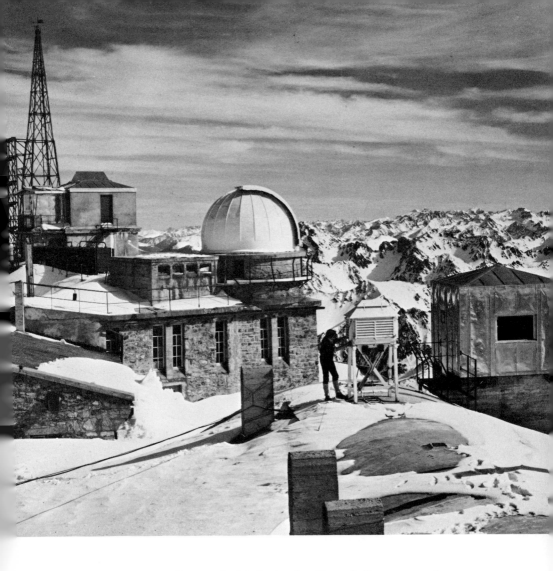

Observatoire du Pic-du-Midi in the French Pyrenees, at an altitude of 9,387 ft.

entire surface of the Moon or Mars) disclosed evidence of any features which could not be accounted for by a gradual work of external processes (such as impacts or meteoritic "weathering")—processes which must demonstrably be operative—while in no case has the existence of processes which on Earth would be called "volcanic" been as yet established. The latter may, to be sure, have also participated in shaping up the surfaces of our nearest celestial neighbors, but probably to a much less perceptible extent.

On the human side, a glance at the data compiled in our Tables II and III reveals that all lunar and planetary research by means of spacecraft has so far been the exclusive domain of the Americans and the Russians—to which no other nation has so far been able to add any direct contribution (and is unlikely to add for several years to come). It can, furthermore, be noted that in all individual feats of lunar space research accomplished so far the Russians had a priority: for theirs was the first lunar fly-by (Luna 1 in January 1959), the first hard-impact (Luna 2 in September 1959) or circumnavigation (Luna 3 in October of the same year); the first soft landing (Luna 9 in February 1966) as well as the first lunar orbiter (Luna 10 in March of the same year). However, the Russian lead-times over the parallel American achievements have been progressively diminishing; and while it has taken more than two years after the Russian firsts for America to fly by the Moon (Ranger 3 in January 1962) or to score a direct hit (Ranger 4 in

April of the same year), for soft landings or injection of spacecraft into circumlunar orbits the Russian lead-times have been reduced to a few months. And what is more important: the American contributions, when they came, were on so massive a scale as to provide the most part of the new evidence that we now possess.

This is, in particular, true of direct photography of the lunar surface; for in comparison with some twenty-two thousand photographs provided by the Rangers, ten thousand frames sent down by Surveyor 1, or several hundred pictures supplied by Orbiters 1 and 2, similar contributions of the Lunas 9 and 13—valuable as they are in other respects—appear rather unimpressive.* And in planetary research (as distinct from an analysis of the particulate contents of the interplanetary space) the successes have so far been largely on the American side; for only in 1966 did the first Russian spacecraft (Venus 2) effect a planetary fly-by; though the Russians can claim credit for the first hard landing (Venus 3) as well as soft landing (Venus 4) on the surface of the veiled planet.

However, the real historical significance of the Russian deep-space probe lies in the fact that they paved the way, and proved the respective feats to be technically possible; their repetition was doubtless stimulated by a knowledge that they can be done. This was especially true at the inception of lunar space work; for without

* Possibly because not all their findings may yet have been made public.

the Russian sputniks from the autumn of 1959 President Kennedy would probably not have proclaimed in 1961 the Moon an official target for the American astronauts in the latter part of this decade; nor would the U. S. Moon program have assumed its present proportions.

The future historian of our subject will no doubt be impressed by the fact that, in order to be able to reach for the Moon in 1959, the Russian work towards the deep-space rockets must have been started prior to 1950, in difficult postwar years when the ravages of the Second World War were far from overcome. At that time the only person who could have authorized this work was Joseph Stalin—that Ivan the Terrible of modern Russian history—and his courageous initiative at that time confronts favorably with the relative complacency with which the dawn of the space age in 1957 was greeted by President Eisenhower of the United States (or, rather, reflects on the initiative and foresight of their respective advisers). Fortunately, the American nation as a whole reacted much more vigorously; and, as a result, by 1968 we are now reaching for the Moon with the human hand.

With the progress already accomplished, the stage is now set for the manned landings as the next obvious step of lunar exploration, and their return to the Earth with samples of lunar surface material to be analyzed in our laboratories. This feat can now be expected with confidence to be accomplished in the next few years. On the American side, the date (contingent on the completion of Saturn C-5 boosters) is now set for 1969; but since

The 250 ft. steerable radio telescope at Jodrell Bank, University of Manchester, used for observations of celestial bodies in the domain of radio frequencies. (Reproduced by courtesy of the Camera Press, London.)

the Russians may possess the necessary boosters already, we may perhaps see them land on the Moon at an earlier date, and thus uphold the priority which they enjoyed so far at all previous stages of lunar exploration. When it occurs, the impact of this feat will doubtless be profound; for the event of establishing human relations with the Moon will remain unique in the history of mankind on this planet. However, from the scientific point of view, the nationality of the first intrepid explorers to set foot on the surface of our nearest celestial neighbor will be largely irrelevant; whoever they are, they will bring down the same results for the edification of the whole of mankind.

Not many men may reach the Moon and return safely within our lifetime. Yet we, the terrestrial plodders, whom age or other accident may have relegated to the role of onlookers in this great scientific episode soon to unfold before our eyes, should rejoice at the thought that after centuries of growth not less than three distinct branches of human technology—rocket propulsion, long-range radio communications, and electronic computer control—matured far enough for their combination to make interplanetary travel possible at this time. This is an accident for which we, the privileged generation, should be duly grateful to our fate; for all our descendents will only learn about this epic story of the history of science from the records we leave behind of events which we witnessed.

6 · Manned Telescopes in Space

The content of the preceding chapter has brought us to the front line of actual accomplishments of the telescopes in space at the present time. However, behind the current work on satellite launching and telemetry an entire small army of other dedicated pioneers of the space age is engaged in planning follow-on missions to be carried out in the years to come. The aim of the present chapter will be to unveil a little the shape of things to come in one aspect of this follow-on work, and give a brief account of the *manned* telescopes in space which are being planned at the present time.

In Chapter IV we gave a brief account of *automated* telescopes which are now (or will soon be) revolving around the Earth on astronomical missions. Why do we wish to man them in the future? The answer is intimately connected with the tasks which we expect these telescopes to perform; and their complexity will increase with the time. As regards the first generation of the orbiting telescopes (such as the OSO's or OGO's) the special tasks for which these spacecraft have been instrumented are very largely of the survey or monitoring type; and these are readily amenable to automation; for almost anything that these spacecraft will see or record will be novel. This need not, however, be wholly the case of the second generation of space telescopes, which may no longer play the role of mere search cameras or survey instruments; but whose tasks may be reformulated in the light of the results furnished by their predecessors; or even of themselves. A requirement that the system be capable of continuous operation over longer periods of time, and also of modification in orbit as the observational program demands, would call for a system which —if automatic—would have to be too highly redundant to provide for long-term reliability, and too complicated to allow self-modification. The alternative approach of simplified design with manned maintenance and modification in orbit appears at present to offer a more realistic means for meeting the design requirements.

In the past few years a rapidly developing technology has provided us with the capability of launching and operating in space large and complex systems, some of which were described in earlier chapters of this book. During the same period it has been demonstrated—by the Mercury and Gemini programs in the United States, together with parallel programs in Russia—that manned operations in space (including extravehicular activity—

The Ranger spacecraft in the laboratory. (By courtesy of the Jet Propulsion Laboratory, California Institute of Technology.)

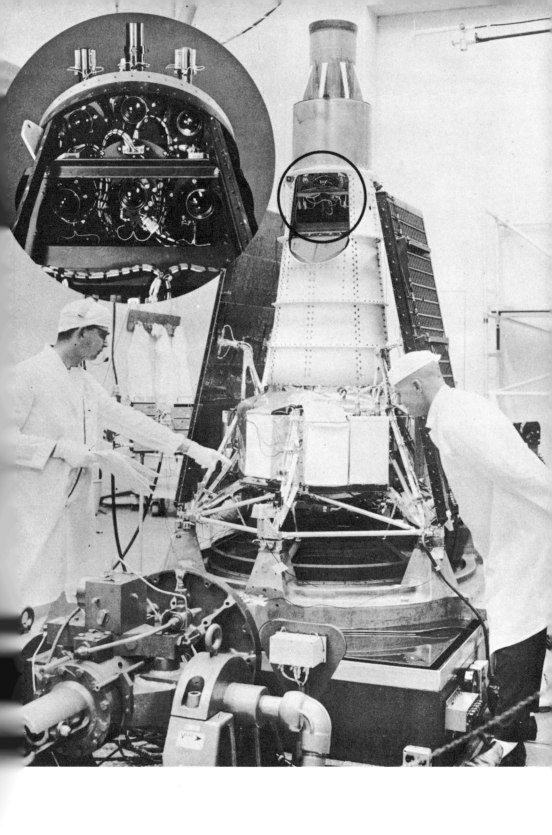

A boring device equipped with a mechanical scoop aboard Surveyor 3 and subsequent spacecraft of this type to test the structure of the lunar surface. The device and its shadow cast on the lunar surface are to the left of the shadow cast by the main mast of the spacecraft. (By courtesy of NASA and JPL.)

(see page 200) can be undertaken with success. Beyond 1970 we can look forward to the development of major manned orbital facilities, in which the astronaut can take part in operating more elaborate equipment and repairing (or replacing) parts that may malfunction; or—more important—maintain liaison by shuttle service with the ground base and keep open a two-way communication system or even the transport of material. When this has been achieved, the stage will be set for a manned orbiting telescope (MOT) which we wish to describe in this chapter (see page 216). Such a telescope is not yet around the corner; but its principles are already under active study; and a pioneer work on its design has recently been completed by the Aerospace Division of the Boeing Company in the United States for the NASA Langley Research Center, proving the general feasibility of the project (see pages 221 and 241). With some luck, ten years from now the entire concept could become an accomplished fact. The aim of the present chapter will be to focus attention to some of the challenging technical problem areas encountered in this connection, as well as to outline briefly the major scientific advances which can be expected to come out from the work with such a telescope.

On doing so, we shall assume that such a telescope would be a modified Cassegrain reflector of Ritchey-Chrétien (or Wynne-Cassegrain) type, with an aperture of the primary mirror close to 120 in.—i.e. about three times as large as that of the largest mirror aboard the

OAO, with ten times its light-gathering power. The range of optical work with such a telescope above the atmosphere should be limited only by available detectors of light, and could well span the entire wavelength range between the domain of hydrogen Lyman series (i.e. about 1,000Å) to the submillimeter waves on the fringe of the domain of radiospectrum—its angular resolution decreasing from 0.01″ at 1,000Å, through 0.05″ in the visible part of the spectrum, to about 1 min. of arc at the millimeter wavelengths.

The speed of work with a telescope depends on the ratio of the aperture of its primary mirror to the effective focal length of the combination; and values of f/15 to f/30 are under consideration for MOT. However, it is possible to generate other focal ratios, for the same combination of mirrors, by purely optical means (an extension by a Barlow lens, or contraction by collimating the beam past a given focus by an auxiliary concave mirror, and rephotographing the field with an auxiliary camera whose f-number then determines the speed of the entire system); so that the adoption of a particular Cassegrain f-number does not predicate that of the entire system.

As to the weight of the proposed telescope, preliminary estimates indicate that an instrument of this size, when assembled on the ground, would weigh approximately 12 tons—in contrast to the 145 tons of weight of the Lick Observatory telescope of the same size. This great reduction in hardware payload is possible because its virtual weightlessness in orbit (or, in more precise terms, the smallness of the gravity gradient torque) permits the whole structure of an orbiting telescope to be very much less massive. Mechanical flexures will form the least part of the design's problems; and it is these that render large ground-based telescopes such cumbersome brutes.

Directly allied with the virtual absence of gravity in orbit, and consequent light structure of the entire telescope, should be its ease of control. The "push-button" acquisition of objects and guiding by means of auxiliary jet motors can be made considerably finer in space than is the case with all ground-based telescopes. As has been proved by supporting work of the Missile Space Division of the General Electric Company (which has also been responsible for the control design of the OAO), automatic guiding (by image-splitting techniques) can be made accurate within errors of the order of 0.01″–0.02″ over prolonged intervals of time; although the "coarse motion" prior to object acquisition by jets will probably be slower than its mechanical equivalent in the terrestrial gravity field.

For the sake of optical performance assessment, let us assume that the quality of the images formed in the focus of the principal mirror of MOT will be limited only by diffraction—which would require that the r.m.s. deviation of its actual shape from the theoretical figure should not be much more than $\lambda/40$ (for about 10 millimicrons for visible light, corresponding to three parts in

10^0 of the mirror diameter). To make such a mirror in the laboratory would be difficult, though not impossible. One reason why this is still a little beyond the present "state of the art" is the fact that such accuracy for any ground-based telescope would be superfluous, and could not really be taken advantage of mainly because of the atmospheric distortion; but for space-borne telescopes freed from atmospheric interference this requirement may become standard.

However, an ability to make such a mirror in the shop is one thing, and to control its shape to the desired accuracy in space is another. Whatever material they are made of—be it glass, or metal, or any other material—mirrors of this size are not entirely rigid and would not maintain their shape to the desired accuracy (on account of the elastic afterworking of residual stresses, solid creep, etc.) unless steps are taken to monitor intermittently their shape in flight and correct any incipient deformations by means of an active servo loop working

The Russian Zond-3 spacecraft, which repeated photography *the far side of the Moon on 28 July 1965 (see page 161). (By courte* *of the USSR Academy of Sciences.)*

on the principle of controlled thermal expansion (activated by suitably placed heat sources) or vacuum deposits beamed in the desired direction, etc. Controlled thermal expansion as a means for touching up optical surfaces of ground-based telescopes has already been investigated with considerable success in France by Couder; and there is but little doubt that the same method may prove to be successful in space as well. In point of fact, without some optical control of this type, large telescopes in space would be seriously incomplete; for it would be obviously pointless to launch a large telescope above the atmosphere whose optical performance would not be limited by diffraction (except for special tasks not requiring high resolution); and no mirror would remain so limited for long periods of time unless its surface could be intermittently touched up by suitable means.

Next, we shall anticipate that the orbiting telescope will be equipped with all auxiliary instruments (spectrographs, photometers, photographic or television cameras, etc.) which we should associate with a similar instrument on the ground (see pages 221 and 241). These alone will, in turn, consist of so many parts requiring a high degree of adjustment, that a completely automatic maintenance would have to be so highly redundant to attain long-term reliability (not to speak of a need for self-modification) as to bog down under its own weight. The maintenance and use of such systems would require a man—in fact, two of them—to inhabit a cabin in which suitable environmental conditions can be maintained, and which orbits with the telescope with which it is connected by soft gimbals. The Boeing engineers, led by D. Bogda-noff, have proved that such a connection can be made not to transmit low-frequency disturbances (such as men moving about in their cabin are likely to generate) and thus not to disturb the high degree of stability of the telescope. That this is indeed possible has been one of the major discoveries of the recent Boeing feasibility study; and it augurs well for the future. As far as the present state of the art of control engineering is con-cerned, nothing can as yet rival (let alone exceed) the storage and programming facility of the human mind in operating the projected telescope, even when the weight of the necessary life-support system is duly taken into account.

The need for having a telescope manned is underlined further by the requirements of data transmission. The results of non-dimensional time studies—such as the photometry of variable stars or other light sources, could —if desirable—be conveniently telemetered to the ground. To a lesser extent, the same should be true of one-dimensional studies, such as the sequential scanning of stellar spectra—but no longer with the transmission of two-dimensional images such as photographs of the star fields. A field half a degree in diameter, imaged with an angular resolution of 0.04″, would consist of 10^9 picture points; and although the current limitations to our image-scanning ability may reduce this number by a factor of

Gemini IV astronaut Edward H. White II walks in space (By courtesy of NASA.)

at least ten, a transmission of 10^8 points would (for the power level delivered by the solar batteries of MOT) occupy all channels of communication for a time of the order of one month.

For the acquisition of such two-dimensional data a direct transfer of material between MOT and its ground bases would be indispensable; and this operation could scarcely be automated, but would require the presence of men aboard. This (plus other logistic aspects of this shuttle service) imposes certain limitations on the orbital altitude at which such a spacecraft can move. In more precise terms, the existence of the van Allen radiation belts surrounding our planet confronts us with a choice between a low orbit (less than 250 miles in altitude) and a high orbit (above 30,000 miles from the Earth). Sustained operations within the van Allen belts would impose a prohibitive penalty in weight needed to protect both men and materials from charged particles and from the X-rays produced by the interaction of such particles with solid matter.

For low orbits, life support as well as transport problems would be easier to handle; but the duration of the "night" aboard a spacecraft revolving in a 2 hr. orbit would not much exceed 40 min. (and the sky above the spacecraft would not be completely black yet, due to the residual auroral activity at higher levels). Moreover, the rapid alternation of "days" and "nights" would entail temperature variations which are still giving some concern to structural engineers. Such environmental disturbances would be greatly lessened for high orbits; the gravity gradient torques—the greatest source of mechanical disturbances—would be 100–1,000 times smaller; and the cyclic thermal changes (which constitute a severe problem for low orbits, due to the spacecraft's periodic occultations by the Earth) become insignificant. Thus the external environmental disturbances are greatly reduced by operating in a high orbit; but, on the other hand, the logistic difficulties of life support and shuttle service between spacecraft and ground are correspondingly increased.

With all these factors in mind, let us outline at least some of the principal gains and advantages in operating a MOT, as envisaged on pages 216, 221, and 241, in space. In the formulation of any observing program it is only natural that the first priority should be assigned for tasks which cannot be approached from ground-based facilities at all—primarily those which require access to invisible parts of the spectrum. This should not mean the total exclusion of observations in the domain of the atmospheric "optical window"; because even for such observations MOT would offer an inestimable advantage of constantly perfect "seeing" at high angular resolution.

This latter capability should underline a choice of tasks for which freedom from atmospheric "bad seeing" may be essential. It would, for instance, be superfluous for most aspects of nebular photography (because of the generally indistinct nature of the objects), or for photo-

Horizontal panorama of a mountainous part of the lunar surface north-east of the crater Tycho, televised by Surveyor 7 in January 1968. (By courtesy of NASA and of JPL.)

electric photometry of single sources (unless the latter happens to be a close double or multiple star). On the other hand, full benefits of highest resolution could be reaped in astrometric or planetary photography, or high-dispersion spectrometry.

The opportunities for unique contributions to astro-nomical research, which could not be approached at all from the ground (or could be done vastly better from space) would tax to capacity not one but several MOT's for many years to come. Of observations of the bodies in the solar system, on the Moon a diffraction-limited 120-inch mirror of MOT should resolve a linear separa-tion of less than 325 ft. in the light of 5,000Å wave-length, and 120 ft. at 2,000Å. On the surface of Venus (observable only for some time around the maximum elongation from the Sun, at a risk of some image degra-dation due to the thermal gradients in optics) the pro-posed MOT would resolve about 9 miles in UV light; while for Mars, Jupiter and Saturn the corresponding resolution at the time of their oppositions should be 5.6, 53 and 109 miles respectively.

A closer view of the lunar crater Copernicus as photographed by the Lunar Orbiter 2 on 24 November 1966. A group of hills in the upper part of the frame constitutes the "Central mountain" of Copernicus; the crater at the bottom is Fauth. (By courtesy of the Boeing Company and of NASA.)

It is true that considerably greater ground resolutions should be attainable even for the outer planets in the 1970s from fly-by or orbiting spacecraft. However, these will be only occasionally operative; or (like the lunar orbiters) will survey only a small part of the surface of the respective celestial body. On the other hand, from the neighborhood of the Earth a global view of the planetary discs would be constantly in front of us, opening new possibilities for the studies of transient (or seasonal) changes on their surfaces as well as of the meteorology (cloud formation) of their atmospheres.

Another important and challenging task of solar-system astronomy should be photoelectric photometry and polarimetry of the occultations of planetary satellites during the atmospheric eclipse stage—a task virtually impossible with ground-based telescopes because of the scattering of the light of the bright planetary disc in our atmosphere. Except at the time of the opposition, the ingress (or egress) of a satellite into (or from) the planet's shadow will occur at a certain angular distance from the planet's limb. The disappearance (or reappearance) of the satellite during such an eclipse (caused by increasing extinction of progressively lower and denser atmospheric layers) is not sudden, but gradual. Monochromatic light curves of such phenomena (observed through filters centered at the principal absorption bands of the planetary spectra) could furnish very important data concerning the stratification of different molecules with the altitude; and any variation in the degree of

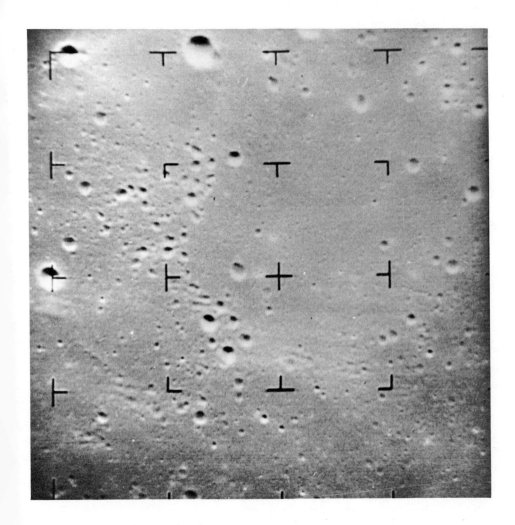

A view of the lunar surface in Mare Cognitum overlaid by a bright ray from the crater Tycho, as televised by Ranger 7 on 31 July 1964 from an altitude of 55 km. The field of view is approximately 26 km across; and the smallest details on the lunar surface are about 50 meters in size. (By courtesy of the Jet Propulsion Laboratory, California Institute of Technology.)

The Russian 236-inch telescope in advanced stage of comple-tion.

polarization of the satellite light during the ingress or egress stage of the atmospheric eclipse would provide important information concerning solid particles (frozen crystals) which may be present at different levels.

The possibilities opening up in stellar astronomy to the users of MOT are so vast that only a brief general description can be given in this place. Freedom from "bad seeing" coupled with diffraction-limited optics should enable the precision of astrometric measurements to be so extended that trigonometric parallaxes could become measurable to a distance of the order of 1,000 parsecs; and thus our entire basis of stellar distance determinations could be placed on a much more secure basis. Sustained positional measurements might also lead to the discovery of new faint companions of near-by stars from the non-linearity of their proper motions.

Another important task of high-resolution photography should concern the investigation of star distribution in central portions of globular (and other dense) clusters, which cannot be resolved from the ground because of the dispersion of light in our atmosphere. Or, turning to photography for astrophysical purposes, a systematic search for very hot stars (supernova remnants; blue subdwarfs, etc.) in the UV; or for nascent cool stars in the IR—spectral domains outside the atmospheric "optical transmission window", can be mentioned as examples of work possible only within severe limitations from the ground. Photographic search for extended gaseous emission in the domain of the hydrogen Lyman series should vastly increase our knowledge of the contents of the interstellar gas.

Possibilities opening up in the spectroscopy (both high- and low-dispersion) of the UV as well as IR domains of the spectrum are beyond easy expression. Moreover, not all spectrographic work should call for complete coverage of wide spectral regions; for as our knowledge increases an increasing number of problems will require detailed surveys of specific parts of the spectrum, and of transient phenomena exhibited by them. This should be true, for example, of most phenomena exhibited by variable stars at different phases of their cycles—in particular, of the periodic line broadening, reversal from absorption to emission, or appearances of emission bands in explosive variables; or the "flash spectra" in the hydrogen Balmer or Lyman lines of eclipsing variables just before and after a total eclipse of their early type components. Such phenomena can best be recorded by scanning spectrographs with photoelectric registration, and (in case of need) telemetered directly to the ground.

An important field of studies to be undertaken by MOT should be concerned with extragalactic objects aiming to extend the present limits of the observable universe by photography, photometry, spectrometry, and polarimetry of the light of distant spiral nebulae and quasi-stellar radio sources. It can be expected with confidence that, in comparison with similar telescopes on the ground, MOT should gain approximately a factor of 0.7 in limiting magnitudes by being above the atmos-

The American Lunar Orbiter spacecraft. (By courtesy of the Boeing Company, Aerospace Division.)

*A horizontal view of the lunar crater Copernicus, as recorded on a
wide-angle photograph taken by the Lunar Orbiter 2 on 24 November 1966. (By courtesy of NASA and of the Boeing Company.)*

pheric extinction, and a further factor of 10 by virtue of better definition of the images unimpaired by atmospheric turbulence. The total gain of magnitude should enable one to suppress the limiting photographic magnitude for detection of giant objects in optical frequencies to +27 and for photoelectric detectors (available so far for registration of point sources rather than of finite areas) this may be further increased to +29.

Under these conditions, direct photography should permit the extension of the present nebular counts to a distance increased by a factor of 10 over that attainable by the 120 in. telescope of the Lick Observatory. Low-dispersion spectroscopy should permit an extension of our present knowledge of the nebular red shift versus apparent magnitude in a similar manner. Photoelectric photometry, with its linear response, should permit the correction of the apparent magnitude scale of distant objects; and also the measurement of the extent of the general reddening of such objects down to the limiting magnitudes attainable with MOT. Such material should, in turn, represent crucial observational checks of various consequences of different cosmological models (explosive, evolutionary, steady-state, etc.) which should permit better discrimination between them than can be so far done from the ground. In doing so, MOT should provide us with data from which to obtain a more meaningful idea of the structure, the past, as well as the future of the Universe in which we live.

The foregoing brief sketch of observing activities

which will become possible when a manned telescope of such power as we envisaged in this book can be launched into a circumterrestrial orbit is very far from complete —as it is difficult to anticipate fuller details years in advance. Once such a telescope (and its successors) become operative, however, a new book of knowledge will be open which has so far been concealed from us by our atmosphere; and its contents may be such as to make our more fortunate descendants speedily relegate most of what we have learned about the Universe from the surface of the Earth into a prehistory of our subject. In point of fact, it does not call for a great exercise of imagination to anticipate that, in the years and decades to come, our ground-based facilities will be gradually relegated into obsolescence as primary sources of astronomical information—these will in the future be located in space (or on the Moon). Who knows how long it may take before our present proud seats of Earth-bound astronomical research may become as relevant to further progress of our science as the fifteenth-century observatory of Ulugh Bek in Samarkand is to modern astrometry today?

7 · Widening Horizons—Space Astronomy of the Future

In the preceding six chapters of this book we have attempted to give a brief outline of the development and potentialities of astronomical telescopes—as well as of the general role they played in the advancement of science—from their discovery in the early part of the seventeenth century to the present time—when telescopes are being launched into space not only to escape the limitations of the atmosphere above us but also (at least within the confines of the solar system) to observe the objects of their investigation at a close range. Where are we likely to go from here?

Astronomy is one of the oldest branches of science conceived by the human mind; and systematic observations of celestial bodies have been conducted from the surface of our planet for more than three thousand years. A retrospective look at the agelong history of our subject reveals its course to be characterized by long periods of slow gestation, separated now and then by sudden efflorescence—when new vistas open up, usually thanks to new tools which the parallel advances in technology have placed from time to time in the astronomer's hands. Such was, for instance, the enchanted period of a few decades following the discovery of the telescope early in the seventeenth century; or the development of the astronomical

An artist's view of the proposed 120-inch Manned Orbiting Tele-scope in space. (By courtesy of the Boeing Company, Aerospace Division.)

reflector at the turn of the eighteenth century, with which William Herschel "coelorum perrupit claustra". The emergence of radio-astronomy in the last thirty years, which unlocked for us the information stored in the microwave domain of the electromagnetic spectrum, led to another sudden expansion of our horizons.

However, none of these landmarks could measure up in their impact and potentialities with the milestone which we are passing today: namely, the advent of space-borne telescopes described in the preceding chapters, the accomplishments of which have ushered astronomy dramatically into the space age. Why? First, by the freedom which they have gained us from atmospheric absorption overhead. Since the dawn of astronomy until about thirty years ago the only channel of optical information and sole link with the Universe around us was visible light, reaching the ground through the relatively narrow atmospheric "optical window" between the violet and the red part of the spectrum; with an additional trickle of radiation reaching us between 8 and 12μ's in the infra-red. These were the days when the old notion that "astronomy is what you see through a telescope" continued to reign supreme.

The advent of radio-astronomy since the late 1930s, which opened up for us on ground the long-wave end of the electromagnetic spectrum, altered somewhat this state of affairs; for radiation at centimeter or meter waves could be "heard" rather than "seen". The main break-through occurred, however, later in our lifetime, when the entire span of the electromagnetic spectra of celestial bodies opened up for inspection of instruments carried above the atmosphere (see page 187). Since the 1950s we are entitled to speak not only of UV-astronomy but also of X-ray or γ-ray astronomy, in much the same way as we speak of optical or radio astronomy. And more than that: we shall soon have to extend the definition of astronomy to cover observations of corpuscular radiation emitted by celestial bodies—of which only neutron or neutrino astronomy can be practiced from the ground;* primary charged particles coming to us from cosmic sources can be observed only well above the terrestrial magnetosphere which represents a much worse obstruction to ground-based particle astronomy than the atmosphere does to light.

In other words, in the future (and, in fact, already now) we shall be hearing of not one, but half a dozen, different "astronomies"—radio, optical, X-ray, etc.—de-

---

* To conduct particle astronomy it is not necessary that the respective celestial body be actually visible to the experimenter. Thus the interiors of the stars can be studied by observations of the neutrinos —ephemeral particles with no charge and next to no mass—which are emitted by certain nuclear reactions going on deep inside the star's core, and can best be observed under heavy material shields. Scientists from Great Britain, India and Japan are currently conducting a search for solar neutrinos at the bottom of a disused gold-mine in India below the terrestrial surface—with results which remain yet to be seen. The results of parallel American experiments carried out by a group of investigators at the California Institute of Technology (Bahcall et al), reported in the summer of 1968, point to a *smaller* flux of neutrinos from the Sun than has been expected—a result which, if significant, may lead to important theoretical repercussions.

pending on the nature of its detectors and channels of information. Moreover, the definition of the telescope itself will have to be broadened to include not only optical or radio telescopes but also the grazing-incidence X-ray, γ-ray, or particle telescopes; the latter being of the nature of pulse-counters rather than image-forming devices. Yet—this is essential—they all will bear out on different observational manifestations of the *same* celestial bodies, and thus will contribute each in its own way to the understanding and reconstruction of the physical processes which are going on in the Universe around us. Specific celestial bodies—such as the well-known Crab Nebula in the constellation of Taurus—have already been observed to emit radiation extending from the radio waves to γ-rays; and also corpuscular radiation.* Only a concerted attack on such sources in all parts of the spectrum—possible only above the atmosphere—may eventually unravel for us the nature of the physical processes which Nature is exhibiting in them on a grand cosmic scale.

Looking further into the future, how deep into space should we plan to penetrate in quest of our scientific objectives? The altitudes necessary for attaining the basic

---

* Needless to say, electromagnetic radiation is bound to remain our principal link with distant celestial bodies, because of the geometric simplicity of its propagation. The vagaries of local interstellar magnetic fields (which play the same role in charged-particle astronomy as do the refraction anomalies for light) give rise to "seeing" so abominable that it is mostly very difficult even to identify the real sources of corpuscular radiation—let alone to measure their positions.

objectives were already mentioned in Chapter 2: a height of some 12 miles above the ground is sufficient to by-pass "bad seeing" and 20–25 miles to open up the infrared part of the spectrum; some 60–90 miles to unlock the near ultraviolet, and 125–185 miles to do the same with extreme UV. Even at such altitudes the sky overhead would, however, not yet be completely black (or, at any rate, as black as we should wish), due to vestigial auroral activity which may reach as far as 600 miles up. It is not till we are well on the way to the Moon that the parochial environment of our atmosphere as well as magnetosphere are safely behind us, and we are really and truly in interplanetary space. What kind of environment should we find ourselves in there?

Unfortunately, we are still not quite out of the woods on our way into really deep space; for, as it was already known to our ancestors, "Nature abhors vacuum", and interplanetary space around the plane of the ecliptic is far from empty. In fact, it contains all kinds of ingredients—ranging from meteorites, meteors, cosmic dust of zodiacal cloud or cometary ices spiraling towards the Sun, with elementary particles of all kinds blown outward by the solar wind—and the sum total of them adds up to a mean density of some 10^{-20} grams per cc. at our distance from the sun, which is about ten thousand times as much as the density prevalent in deep interstellar space. At a distance of a mere 93 million miles from our central luminary, we still live in a very real sense in the extension of the Sun and its outer layers; so

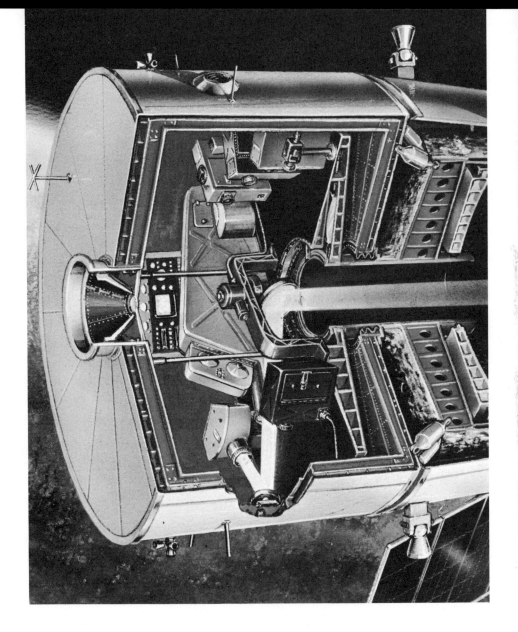

A cross-section of the observing cage of the proposed 120-inch manned orbiting telescope (MOT), showing distribution of auxiliary instruments. (By courtesy of the Boeing Company, Aerospace Division.)

gradually does a star like it peter out into interstellar space.

This zodiacal haze possesses, however, one characteristic feature enforced by solar attraction: namely, it is largely concentrated towards the plane of the ecliptic in the vicinity of which most planetary bodies revolve. This suggests the possibility of getting out of this haze in a direction perpendicular to the ecliptic. A spacecraft launched normally to a direction to the Sun, in an orbit of the dimensions of those of the Mariners or other interplanetary spacecraft already launched, should be able to emerge near its apogee well from the ecliptical fog and have a good look through its telescopes into really deep space. How far could we see into it from this vantage-point?

The reader who would expect that, at this distance from our terrestrial abode, we should be completely free from any environmental limitations is again likely to be disappointed; for nature continues to abhor vacuum wherever it can find it, and filled also interstellar space with a tenuous gas of mainly neutral hydrogen. This gas is so tenuous that, on the average, not more than one to ten hydrogen atoms are found in each cubic centimeter. But, on the other hand, the dimensions of our galaxy and expanse of interstellar space is so vast that enough

Orbiting Astronomical Observatory (OAO)

Orbiting Geophysical
Observatory (OGO)

Orbiting Solar Observatory (OSO)

Orbiting observatories.

such atoms may be encountered by a ray of light along the line of sight if we go far enough to produce undesirable optical effects; and these may—by analogy with our atmosphere—give rise to concern about interstellar absorption and "seeing".

The latter—namely, the fear of poor "interstellar seeing"—with the wave-front becoming corrugated as a result of inhomogeneities in interstellar gas through which the light passes—can be dismissed out of hand; for since the mean free path of all particles in interstellar space amounts to many millions of miles, any refraction effects caused by local interstellar turbulence should be uniform across the aperture of any conceivable telescope or interferometer. In brief, any imperfections in interstellar seeing, giving rise to scintillation of interstellar origin, are probably too small to be detected—at least at optical frequencies.

However, the same is not true of the emission or scattering of light which interstellar hydrogen gas can exert at long distances. In particular, in the extreme ultraviolet —at wavelengths between the Lyman limit ($\lambda = 912\text{Å}$) of the hydrogen atom and the domain of soft X-rays (from $\lambda = 50\text{Å}$ up)—interstellar space too is apt to be decidedly "hazy", with visibility rather seriously reduced —particularly along the galactic equator, where most of interstellar hydrogen is concentrated. In order to penetrate through it to distant parts of our Milky Way, or into intergalactic space, our spacecraft will have to be directed to fish for a glimpse through hydrogen-free "galactic windows" (which are known to exist even close to the galactic equator), or direct their sights towards high galactic latitudes, where not so many hydrogen atoms are likely to obstruct our view. About limitations to transparency of the intergalactic space we can as yet say nothing; nor are we likely to need this information in space astronomy for many years to come.

In summary we wish, however, to re-emphasize the fact that a deep penetration of the galactic space is possible only on wavelengths which are very different from those of visible light. Two kinds of obstacles are in the way of light waves from distant parts of the galaxy—and in particular, from the direction of the galactic center where most part of the mass of our Milky Way system is concentrated: namely, interstellar gas and dust. It is particularly only the latter which deprives our eyes of a grand spectacle of the central bulge of our galaxy, which would otherwise dominate our southern sky in the direction of the constellation of Sagittarius. The interstellar haze obscuring it from view to visual observations is so dense in this direction that the optical outlines of this bulge did not begin to emerge from the sky background in deep infrared till long after its location had been determined from dynamical considerations.

In principle, thc combined interstellar haze of gas and dust becomes transparent in two widely separated domains of the electromagnetic spectrum: namely, the X-ray domain and that of radio-frequencies. The latter (comprising wavelengths longer than approximately 1 mm.) possesses the advantage of penetration through our atmosphere, and allows observations from ground-based facilities; while observations in the X-ray domain can be made only from space. However, the optical resolution for a given aperture is at its worst for radio-frequencies, and becomes best for the shortest wavelengths which can be focused with the telescope. In other words, the methods of radio-astronomy can enable us to discern the objects obscured by interstellar medium from ground-based observatories, but only vaguely so, because of severely limited angular resolution; and consequently, any identifications of radio sources with luminous objects is always uncertain (unless helped by interferometric techniques). On the other hand, in the X-ray domain angular resolution is potentially much superior to that in visible light; but can be realized only in space, and its exploitation will call for means not yet in practical use.

But after this excursion to the frontiers of our galaxy, let us return to our own station within it, and consider briefly another problem of transcending significance whose solution can be advanced only by the methods of space astronomy: namely, is there life elsewhere in the Universe than on the Earth, and how to find out about it? This quest is providing indeed the main impetus behind the impending exploration of Mars; and the results already furnished by Mariner 4 revealed that such life as may be vegetating on the surface of the red planet will scarcely be of interest to anybody outside the ranks of professional biologists.

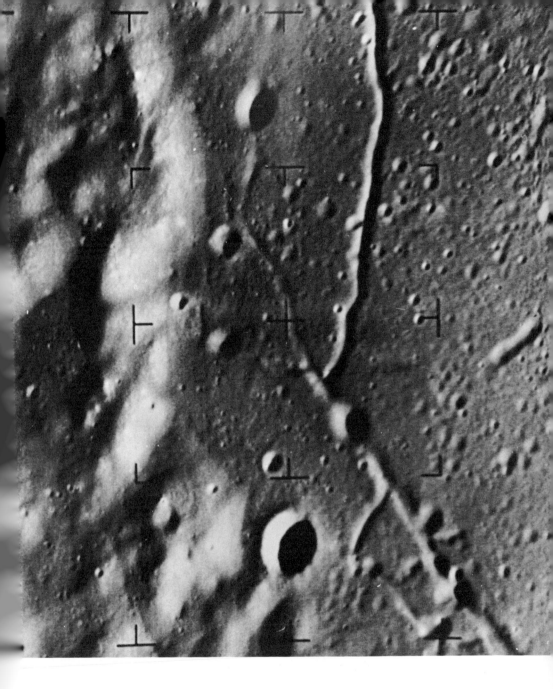

A part of the floor of the lunar crater Alphonsus, as televised by Ranger 9 on 24 March 1965 from an altitude of 184 km above the lunar surface. A section of the walls of this crater can be seen on the left. (By courtesy of the Jet Propulsion Laboratory, California Institute of Technology.)

But how about outside our solar system? The possibility of its existence hinges, of course, on the discovery of whether or not any other stars within observational reach possess planetary systems; for life as we know it on Earth can develop and flourish only on a planet located at the right distance from its central luminary for water to exist on its surface in liquid form; and also of the right mass: for if it were too massive its atmosphere could have retained hydrogen compounds which are lethal to life as we know it; while if its mass were too small it would not retain any atmosphere (and, thus, liquid on the surface) at all.

Before we have any need to worry about these aspects it is, of course, necessary to discover first whether any other star around us possesses a planetary system of its own; for only then the question of its suitability for life support can really arise. In the absence of any positive knowledge of the origin of planetary systems, the best strategy may be to attack the problem from the observational side.

The principal difficulty in discovering planets in the proximity of their central stars is, of course, the great disparity in brightness of these two classes of celestial bodies; and a combination of this difference with proximity makes any observational discovery of such systems utterly impossible by any kind of equipment used on the ground. Space astronomy offers obviously the only avenue of approach; but even then observational difficulties are still formidable.

In order to appreciate them, consider the converse problem of discovering the existence of Jupiter—the largest planet of our solar system—from our nearest stellar neighbors in space a few parsecs away. At the same distance, our Sun would appear to an external observer to be about two milliard times as bright as Jupiter in visible light—corresponding to a difference of approximately twenty-three stellar magnitudes; and although this latter difference could be diminished to some eleven magnitudes for observations carried out sufficiently far in the infrared (for $\lambda \sim 30\mu$), the concomittant loss of resolving power for a given aperture (together with reduced quantum efficiency of the requisite IR detectors) would make it desirable for us to confine our work to the optical domain. For at the distance of a mere 5 parsecs (within which not more than thirty-five stars in our neighborhood are located) the angular separation between them would be only 1.1 sec. of arc (0.2″ for the Earth) and proportionally less at greater distances. How do we discover such close pairs of light sources so grossly unequal in brightness?

One possibility which suggests itself in space is to employ a suitable occulting disc to block off the light of the central star. However, in order to minimize diffraction of light on the edge of such a disc, the latter would have to be placed at a considerable distance from the primary optics. A recent closer analysis of the situation by R. Danielson at Princeton revealed that, for a space telescope of approximately 1,000-inch of free aperture,

*The American Mariner Spacecraft—an eerie weird bird which tra-
versed first the distance to Venus.*

an occulting disc of 75 m. in diameter held at a distance of 6,200 miles would be adequate to detect a planet separated from its star by 1″ of angular distance; and a 1,000 ft. obstacle at a distance of a quarter of a million miles should permit us to increase the angular resolution to 0.2″.

The large distances needed to separate the occulting discs and the optics would greatly complicate control of the experiment; since the relative motion across the line of sight would have to be kept very small. Observation of the disc from the telescope—perhaps a flashing light —would provide an error signal which could be transmitted back to a servomechanism controlling the obstacle. At a distance of 6,000 miles precise control might be possible with moderate power in a circumterrestrial orbit; while at a separation of 155,000 miles or so a heliocentric orbit for the occulting disc may be more appropriate.

Should this method fail to prove practicable, one might attempt to detect the existence of other planetary systems by a more indirect method—proposed by the Russian astronomer Fessenkov in 1961—based on a search for the "zodiacal light" around a star, which is likely to accompany each planetary system, and whose particles represent probably the left-overs from the time of its formation. The total amount of sunlight scattered by our own zodiacal cloud exceeds many times that of all planets; so that the disparity in luminosity between the central star and the objects sought after around it would not be so overwhelming. Moreover, the direction in which a "zodiacal cloud" around another star would appear to be elongated could indicate to us the orientation of the "invariable plane" around which its principal planets revolve; and thus the locus in which their direct optical images should be sought.

On the other hand, although the total brightness of such a cloud is likely to exceed that of all planets, its surface brightness (per unit area) may again be too low for observational detection; and the discrimination against scattered or diffracted light from the central star might well prove to be as difficult as the detection of individual planetary images. The American astronomer Spitzer conjured up in 1962 the vision of a hypothetical telescope for this purpose, 37 miles in aperture, the mirror of which could be carved out of an asteroid; but he noted that solar tides alone would distort its shape by intolerable amounts if it were operated anywhere within the orbit of Pluto.

In the case of these facts, it may well prove more feasible, in the future, to send out instrumented probes on a reconnaissance expedition in the neighborhood of the near-by stars than to build the equipment necessary to search for their planets with adequate resolution from the proximity of the Sun. The reader may recall from previous chapters of this book that a 4-inch lens of a lunar Orbiter did, from its vantage-point, resolve on the lunar surface details smaller than those which could be observed by a 120-inch telescope orbiting around the

Earth. And the same may prove to be true, at a later stage of space exploration, of the investigation of other stars (or of the search for their planets) as well.

At the present time no spacecraft powered by chemical propellants can escape from the gravitational confines of our solar system. However, with nuclear rocket propulsion around the corner (astronomically speaking), this may become another story; and ion rockets of the future should be able to accelerate dirigible spacecraft to velocities which may not fall too far short of that of light. When this becomes possible another great milestone in the development of space exploration will have been reached; for from then on direct exploration of celestial objects at a close range will no longer be limited to the solar system, but can be extended to the realm of the stars.

There are altogether thirty-five stars in our proximity up to a distance of 5 parsecs (16.3 light-years); and over one hundred are known to be closer to us than 10 parsecs (32.6 light-years). Once ionic rockets become operative, there is nothing improbable for a spacecraft to spend years on an interstellar mission, and telemeter to us the results of its reconnaissance; but which technical means can be used for this transmission? Obviously not the conventional radio-waves or even microwaves; for their dilution across the intervening abyss of space would be so tremendous as to require inordinately long transmission times (or tremendous power for the transmitter). Remember in this connection, that a dilution factor of the order of 10^{-16} from the distance of Mars necessitated transmission times of eight hours for a single picture—how much longer these would be from the distance of a star!

Fortunately, we already possess technical means which may shorten these transmission times to reasonable limits: and that is by the use of *laser*. The term itself (an acronym for light amplifier with stimulated emission of radiation) calls perhaps for a few words of explanation; and the best way to do so will be by comparing the essential properties of laser beams with those of ordinary light. As is well known, ordinary white light is produced by electromagnetic vibrations in arbitrary planes, with arbitrary phases, and arbitrary mixtures of wavelengths. It is in general possible to lessen the mixture of wavelengths by appropriate filters or other devices; and atomic processes are known to emit light which is pretty nearly monochromatic (i.e. in which all vibrations possess the same wavelengths). It is, furthermore, possible by means of suitable devices to segregate waves vibrating in the same plane (and thus obtain light which we call polar-

A model of the Russian space station Luna 3, which on 7 Octobe 1959 secured the first photographs of the Moon's far side (see pag 150). (By courtesy of USSR Academy of Sciences.)

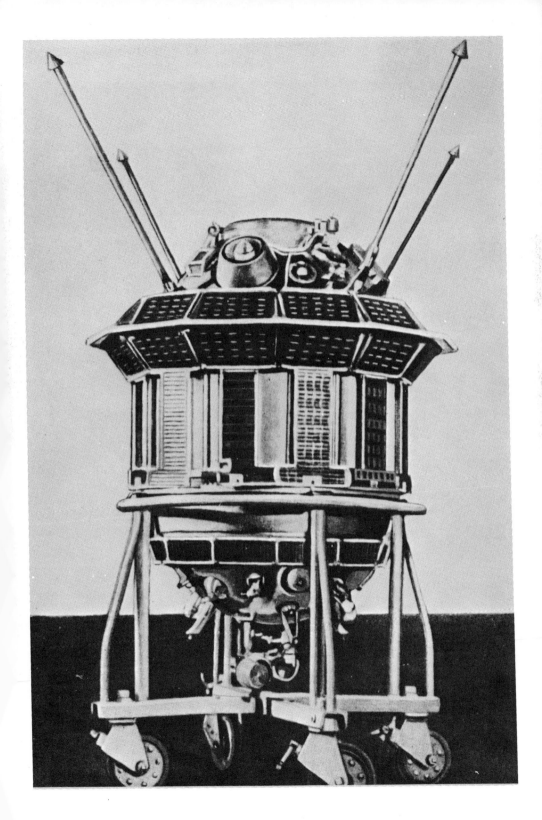

Surveyors 5-7 carried to the Moon this chemical sampler (resting on the lunar surface) which can reveal the atomic composition of the lunar soil from its α-particle scattering properties. (By courtesy of NASA and JPL.)

ized); but even in such a beam the phases of individual vibrations may still be entirely uncorrelated (and thus constitute light which lacks coherence).

Sunlight, or artificial light of electric bulbs, fluorescent tubes and many other well-known sources, are not coherent, because the atoms which emit their light do so independently of each other. On the other hand, at lower frequencies, electromagnetic waves transmitting radio or television are coherent because the electrons whose oscillations are separately responsible for producing the radio energy are made to move in concert with each other within the appropriate electrical circuits. However, coherent light in optical frequencies—the laser—was not excited by human hand until in the last few years.

The essential idea of any laser is to ensure that excited atoms of suitable paramagnetic substance (such as the crystals of synthetic ruby, gallium arsenide, or certain inert gases) in a magnetic field will emit their light quanta, not haphazardly, but in unison. This can indeed be achieved by bathing them in a radiation capable of inducing a chain reaction: a single pulse of light, emitted by a single atom, will emerge from the laser material as a shower. In the ruby and gas lasers this shower of pulses can be amplified into cloudburst by making the light bounce several times between parallel mirrors, so as to traverse the breadth of the material more than a hundred times, and be intensified each time.

As a result, the intensity of the light flashes emitted by lasers can be truly enormous. One flash of this kind given out by a ruby crystal can produce one calorie (i.e. about 10^7 ergs) of energy in 10^{-8} scc. Unfortunately, this flow cannot as yet be sustained for much longer intervals of time; but while it lasts, one small ruby crystal can emit energy at a rate comparable with that of the largest existing nuclear power station. Indeed, laser constitutes the most efficient way for squeezing out energy from atoms without disturbing their nuclei.

Although the coherence as well as strictly monochromatic nature of laser beams can make them eminently useful in many branches of science, so far they have been used mainly for the intensity of light which they can produce; and the first spectacular proof of its power has been the successful attempt of the physicists Smullin and Fiocco from the Lincoln Laboratories of the Massachusetts Institute of Technology in the United States in May 1962 to observe the reflection of its light waves from the Moon. In this pioneer experiment a ruby laser was made to send out pulses of half a millisecond duration at the wavelength of 6,943Å (i.e. in the red part of the spectrum). The light of a flash repeated each minute, and collimated by a 12-inch mirror, formed so parallel a beam that the area of the Moon's surface illuminated by it was not more than 2–3 miles across. The light reflected back was naturally too faint to be visible by the naked eye; and in order to detect it a red photomultiplier had to be placed in the focuse of a telescope of 50-inch free aperture. Out of some 2×10^{23} photons sent out to the Moon, the telescope recorded the return of only about twelve, after a round-trip of some 477,000 miles accomplished in 2.56 sec. The fact that the few returning photons could be identified we owe entirely to their coherence—just as the same property of radio-waves enables us to identify the signals sent out from deep-space probes with energies of a few watts at distances of tens of hundreds of millions of miles.

In the near future we shall see laser beams transmitting telephonic conversations on Earth in much the same way (but much more efficiently) as our present transcontinental UHF links via the Early Bird satellites—though at present the only place on the Earth where enough telephone calls are being made within sight of both parties to justify this installation would seem to be Manhattan, New York. There is no doubt that, in the years to come, they will replace altogether all radio links between rockets in space. It is only in space-to-ground communications that a laser link is handicapped by indifferent transparency of our atmosphere; for laser light does not penetrate clouds or fog any better than ordinary visible light of the same wavelength. It is probable that, in the future, laser messages from spacecraft will be monitored by observing platforms on the Moon or in orbit around the Earth, and transmitted to the ground at radio frequencies which penetrate through both the ionosphere and the air below. In this way—using coherent light in the visible (or, better, X-ray) domain of the spectrum for signal transmission—we should find it possible to take advantage of space transparency to transmit intelligible messages that should be reasonably free from noise even across abysmal distances which separate us from the stars.*

The importance of this channel of information is underlined by the fact that—unlike ordinary light—coherent light is not produced by any natural processes known to physical sciences. Therefore, should we discover any such sources in the sky—especially pulsed sources, or those exhibiting any distinguishable pattern—the conclusion would be inevitable that they were produced by intelligent action of other beings.

A watch for coherent messages from the Universe in the domain of radio-frequencies has already been started in the last decade in the United States (Project OZMA); and the fact that it has so far been fruitless should not detract from its potentially high interest; for in a matter as important as a search for intelligent life on other celestial bodies one cannot expect to obtain positive results in a few years. The principal drawback of radio-astronomical approach to the problem is, of course, the low angular resolution of the underlying observations. Suppose, for the sake of argument, that we do detect a source of coherent radio emission in the sky: the "field

---

* We should, of course, bear in mind that the signals sent out from observing platforms that move through space with velocities close to that of light will be enormously displaced in frequency because of the Doppler effect. Thus, on the outward journey, X-ray signals might be made to drift into the domain of visible light; and vice versa after circumnavigation of the object—a fact of which advantage could be made in experiment design.

A cross-section of the proposed 120-inch manned orbiting telescope (MOT) in space. (By courtesy of the Boeing Company, Aerospace Division.)

of view" (i.e. beamwidth) of the radio-telescope employed would probably encompass hundreds of individual stars which one can photograph—and at very different distances from us—so that a unique identification of the source would be virtually impossible (unless, perhaps, interferometric techniques could be employed).

That this should be clear to the originators of such signals we can anticipate with confidence if they are at least equal to us in intelligence; for to signal their position by radio would be of only a limited help for identification. If, however, they are more intelligent than ourselves, and can produce coherent beams of X-rays of sufficient intensity, they should not only be able to reach a similar galaxy-wide audience, but would also provide adequate means for pin-pointing the source; and, thus, the likelihood that their message could be acknowledged and returned from other parts of the Galaxy would be correspondingly increased. Are, by any chance, any such messages being beamed on us while I write or you read these pages? We cannot say; for we lack as yet any systematic program of watching for them on the ground, because X-rays do not penetrate our atmosphere. For this we shall have to await the establishment of observing bases in space or on the Moon—hopefully in the not too distant future.

And with this prospect we have come to the end of our few glimpses of the shape of things to come in the field of telescopic space exploration and of astronomical science in general. In the sketch which I endeavored to give of this fascinating subject I have led my reader to the very confines of our present knowledge. It is barely ten years since the first spacecraft made by human hand was launched to defy the law of free fall, to be followed by hundreds of others from terrestrial to heliocentric orbits. In the preceding chapters of this book we surveyed briefly what has already been discovered with their aid, and there seems scarcely discernible limit to what may be discovered ten—let alone a hundred—years from now. Some of the anticipations which we ventured to set forth in this chapter may perhaps then be shown to be false; but it is profoundly improbable that we have been oversanguine in our expectations—if anything, the opposite will probably be true; for the actual Book of Nature is much more exciting than any anticipations.

TABLE I

Orbiting Observations

Initial Orbital Data

NAME	DATE OF LAUNCHING	WEIGHT (LB.)	PERIOD (MIN.)	PERI-GEE (MI.)	APO-GEE (MI.)	IN-CLINE (DEG.)	REMARKS
Sputnik 3	May 15, 1958	2,925	105.8	140	1,166	62.5	Decayed April 6, 1960
OSO-1	March 7, 1962	458	96.2	343	367	32.8	Transmitted data until Aug. 6, 1963
OGO-1	Sept. 4, 1964	1,073	3840	175	92,622	31.1	Spin stabilized; in orbit
OSO-2	Feb. 3, 1965	545	96.5	342	392	32.9	In orbit
OGO-2	Oct. 14, 1965	1,118	104.3	259	939	87.4	In orbit
OAO-1	April 8, 1966	3,900	100.8	450	505	35.0	Battery failed 2nd day in orbit
OGO-3	June 6, 1966	1,135	2908	170	75,600	30.9	In orbit
OSO-3	March 8, 1967	627	95.9	335	353	32.9	In orbit
OGO-4	July 28, 1967	1,216	98.1	255	562	86.0	In orbit
OSO-4	Oct. 18, 1967	597	95.7	333	353	32.9	In orbit
OGO-5	March 4, 1968	1,347	3537	180	90,973	88	In orbit
OAO-2	Dec. 7, 1968	4,446	100	479	485	35.0	In Orbit
OSO-5	Jan. 22, 1969	641	95.6	333	349	32.9	In Orbit

TABLE II

NAME	ORIGIN	DATE OF LAUNCHING
(A) FLY-BY		
Luna 1	USSR	January 2, 1959
Luna 3	USSR	October 4, 1959
Ranger 3	USA	January 26, 1962
Ranger 5	USA	October 18, 1962
Luna 4	USSR	April 2, 1963
Luna 6	USSR	June 8, 1965
Zond 3	USSR	July 18, 1965
(B) HARD-LANDING		
Luna 2	USSR	September 12, 1959
Ranger 4	USA	April 23, 1962
Ranger 6	USA	January 30, 1964
Ranger 7	USA	July 28, 1964
Ranger 8	USA	February 17, 1965
Ranger 9	USA	March 21, 1965
Luna 5	USSR	May 9, 1965
Luna 7	USSR	October 4, 1965
Luna 8	USSR	December 3, 1965
Surveyor 2	USA	September 20, 1966
Surveyor 4	USA	July 14, 1967

(C) SOFT-LANDING

Luna 9	USSR	January 31, 1966
Surveyor 1	USA	May 30, 1966
Luna 13	USSR	December 21, 1966
Surveyor 3	USA	April 17, 1967
Surveyor 5	USA	September 8, 1967
Surveyor 6	USA	November 7, 1967
Surveyor 7	USA	January 7, 1968

(D) ORBITING

Luna 10	USSR	March 31, 1966
Orbiter 1	USA	August 10, 1966
Luna 11	USSR	August 24, 1966
Luna 12	USSR	October 22, 1966
Orbiter 2	USA	November 6, 1966
Orbiter 3	USA	February 5, 1967
Orbiter 4	USA	May 4, 1967
Explorer 35	USA	July 19, 1967
Orbiter 5	USA	August 1, 1967
Luna 14	USSR	April 7, 1968
Luna 15	USSR	July 13, 1969

TABLE III

Planetary Probes

NAME	ORIGIN	DATE OF LAUNCHING	DESTINATION
Venus 1	USSR	February 12, 1961	Venus (contact lost after 4.7 million miles)
Mariner 2	USA	August 26, 1962	Venus (fly-by on December 14, 1962)
Mars 1	USSR	November 1, 1962	Mars (contact lost after 65.9 million miles)
Zond 1	USSR	April 2, 1964	Venus (contact lost en route)
Mariner 3	USA	November 5, 1964	Mars (failed by shroud malfunctioning)
Mariner 4	USA	November 28, 1964	Mars (fly-by on July 14, 1965)
Zond 2	USSR	November 30, 1964	Mars (batteries failed after May 5, 1965)
Venus 2	USSR	November 12, 1965	Venus (fly-by on February 27, 1966)
Venus 3	USSR	November 16, 1965	Venus (crash-landing on March 1, 1966)
Venus 4	USSR	June 12, 1967	Venus (landing on October 18, 1967)
Mariner 5	USA	June 14, 1967	Venus (fly-by on October 19, 1967)
Venus 5	USSR	January 5, 1969	Venus (landing on May 16, 1969)
Venus 6	USSR	January 10, 1969	Venus (landing on May 17, 1969)
Mariner 6	USA	February 24, 1969	Mars (fly-by on July 31, 1969)
Mariner 7	USA	March 27, 1969	Mars (fly-by on August 5, 1969)

Index